CHOW!

优雅的中餐

周德丽 —— 著
黄国雄 —— 绘

广西师范大学出版社
·桂林·

CHOW!
Copyright © 2017 Cecilia Carolyn Hsu
with the support of the Mattawin Company
New and completely revised edition of the book originally published in Shanghai in 1936.

图书在版编目(CIP)数据

优雅的中餐：中文、英文/周德丽著；黄国雄绘.—桂林：广西师范大学出版社，2019.4(2022.1 重印)
ISBN 978-7-5598-0853-0

Ⅰ.①优… Ⅱ.①周… ②黄… Ⅲ.①饮食－文化－中国－英、汉 Ⅳ.①TS971.202

中国版本图书馆 CIP 数据核字(2018)第 092904 号

出 品 人：刘广汉
策划编辑：尹晓冬　宋书晔
责任编辑：刘孝霞
助理编辑：宋书晔
装帧设计：李婷婷

广西师范大学出版社出版发行
(广西桂林市五里店路9号　　邮政编码：541004)
(网址：http://www.bbtpress.com)
出版人：黄轩庄
全国新华书店经销
销售热线：021-65200318　021-31260822-898
山东韵杰文化科技有限公司印刷
(山东省淄博市桓台县桓台大道西首　邮政编码：256401)
开本：787mm×1 092mm　1/32
印张：11　　　　　　　字数：201 千字
2019 年 4 月第 1 版　　2022 年 1 月第 2 次印刷
定价：58.00 元

如发现印装质量问题，影响阅读，请与出版社发行部门联系调换。

CONTENTS

Pages

Preface for the Reprinted Version	1
Introduction	9
Preface	13
Foreword	17
The Housewife and Cooking	1
The Art of Cooking	6
Methods of Cooking	8
Flavouring	13
Use of Flour	14
Selection of Ingredients	14
Serving	15

Dinner Parties	17
Restaurant Dinners	18
Home Dinners	19
Plain Daily Meals	20
Table Manners	37
Seating Arrangements	38
Don'ts at a Chinese Dinner Table	45
How to Hold Chopsticks	46
How to Hold Rice Bowls	48
Table Service	57
Tea	60
Wine and Song	64
Kitchen Utensils	70
Ingredients and Condiments	73
Soya Sauce	73
Shaohsing Wine	73
Winter Mushrooms	74
How to Boil Mushrooms	74
Red Haws	76
Bamboo Shoots	76
Ginger	76
Chinese Onion	76

Selected Recipes

Basics	83
Fowl Dishes	92
Meat Dishes	132
Seafood Dishes	164
Vegetable Dishes	202
Soups	228
Desserts	246
Pastries	260
Suggested Menus	295
Index for Chinese Words	308
Index	310
Afterword	317

目 录

- 5 再版序言
- 11 序一
- 15 序二
- 21 自序

- 22 家庭主妇与烹饪

- 25 **烹饪的艺术**
- 26 烹饪之方法
- 29 调味法
- 29 面粉之用法
- 30 选料
- 30 侍餐

- 32 **宴席**
- 33 酒席
- 34 家厨酒席
- 35 家常便饭

- 51 **餐桌礼仪**
- 52 座位安排
- 54 中国餐仪之禁
- 55 执箸
- 56 持碗

- 59 **餐桌食具**
- 62 茶
- 68 酒与歌

72	**炊具**
78	**配料与调料**
78	酱油
78	绍兴酒
78	冬菇
79	蘑菇汤
79	山楂饼
79	竹笋
80	姜
80	葱

中馈珍藏

83	**基础类**
93	**禽类**
133	**肉类**
165	**河海鲜类**
203	**菜蔬类**
229	**汤类**
247	**甜点心**
261	**面食**
302	**建议菜单**
314	**索引**
319	**后记**

PREFACE FOR THE REPRINTED VERSION

Great-grand-aunt Dolly Chow was known for being a larger-than-life fun-loving lady. She was raised in Shanghai where her father Sir Chow Shouson (周寿臣爵士), a member of the first Chinese Educational Mission to the West and a prominent trade official during the Qing dynasty, served the nation in a number of government posts. In addition to being attentive to a busy husband [C.T. Wang (王正廷), an international diplomat, community leader and father of our Chinese Olympics] and overseeing a household, great-grand-aunt Dolly herself had great culinary skills and was a gracious hostess. An invitation to dinner at the Wang household was highly appreciated.

Great-grand-aunt Dolly decided to commit her culinary knowledge to paper, and so she infused her family's private banquet menu with traditional Chinese table etiquette and drafted it all into a book entitled "Chinese Recipes" (《中国食谱》)[1]. Great-grand-aunt Dolly and great-grand-uncle C.T. had many western friends and Dolly wanted to share our Chinese culture with them by introducing them to the philosophy of traditional Chinese dining and etiquette. This was her main purpose for writing this book. She observed that the common preference among her international guests was to taste various types of Chinese cooking. Therefore, great-grand-aunt Dolly's book was written in both Chinese and English for everyone to enjoy. First published in 1936 in Shanghai then reprinted 68 times in Shanghai, Hong Kong, Tokyo and New York, the book has sold over 300,000 copies.

Today Chinese customs and practices are not such a mystery to the West as they were in great-grand-aunt Dolly's day. However we feel a sense of mission driving us to republish the book. As Chinese society has undergone extreme changes, in terms of economic growth and social transformation over the last 70 years, many young people have forgotten or have never learned that like westerners, we Chinese have a tradition of table etiquette and manners. Young people tend to think that table etiquette and manners are derived and imported from western civilization. Far from it.

Our young Chinese generation would gain by remembering

1　When first published, the book was named " Chinese Recipes " . Editor's note.

the teachings of humility and grace passed down from the social teachings of Confucius, teachings that deal with the proper behavior and manners of the individual in society. Confucius had many rules on dining etiquette as he believed the function of eating was not only a way to satisfy one's appetite but was also the most important means of social interaction. Chinese table etiquette and manners may seem unremarkable, but they are indeed profound.

I have often shared this book with my young Chinese friends in both China and abroad. It is my way of inspiring others to read and learn more about our Chinese heritage, to be more knowledgeable about our social history, to understand the deep roots that table etiquette and manners have in Chinese culture and to be proud of our culture. Just as importantly, I believe this book will give young people important advantages as they progress through society and interact with people from a wide variety of cultures. To that end, I have added commentary to select chapters that helps bridge the gap between the 1930s and today.

In all cultures, a shared meal is at the centre of life's most important moments. Whether dining for business or pleasure, whether meeting one's spouse for the first time or one's future boss, proper manners and etiquette will carry one a long way.

This book has been out of print for a long time and can hardly be found today. Thanks to the support of Guangxi Normal

University Press, this book is being republished. Thanks to Mr. Xu Zhongliang for providing his support. Mr. Xu also engaged Ms. Liu Yanqi who formerly was a United Nations translator to translate parts of the book from English to Chinese. The layout and illustrations have been kept intact from the original edition and enhance the book's traditional charm.

Carolyn Hsu 徐芝韵
New York, June 2018

再版序言

我的太舅婆周德丽（Dolly Chow）是一位热爱生活的卓越女士。她在上海长大，父亲周寿臣爵士（Sir Chow Shouson）是清政府首批派童赴美的留学生之一，后在清政府担任重要贸易官员，并在中华民国时期担任一系列政府职务。除了悉心照料公务繁忙的太舅公［王正廷（C.T. Wang），外交家、社区领袖和中国奥林匹克之父］，照管家庭上下事务以外，太舅婆还非常擅长烹饪艺术，是一位亲切随和的女主人。受邀到她府上参加家宴是宾客的莫大荣幸。

太舅婆决定将她所谙熟的中国传统美食知识写下来，于是便将她家的私家家宴菜谱和传统的中餐餐桌礼仪融会于此，著成了《中国食谱》[1]一书。太舅婆与太舅公有很多西方朋友，她希望借由展示中国传统美食和餐桌礼仪之道，向他们介绍中国文化，这是她写作此书的主要目的。她发现，国际宾客们的一个共同喜好是品尝各种中国美味佳肴，因此便以中英文写就此书，方便各国读者阅读、欣赏。这本书于1936年在上海初次出版，之后在上海、香港、东京和纽约等地再版六十八次，销量逾三十万册。

今天，中国人的传统习惯对西方国家来说虽然已不再像

---

1 本书初次出版时中文名为《中国食谱》。——编者注。

太舅婆他们那个时代那样神秘，但有种使命感促使我们再次出版此书。中国社会过去七十多年来在经济增长和社会转型方面经历了巨变，很多年轻人已忘记或从未了解我们中国人和西方人一样，也有优雅的传统餐桌礼仪，这往往导致他们认为餐桌礼仪源自西方文明。事实远非如此。

中国年轻一代将从有关谦卑和感恩的传统教诲中受益匪浅，这些源自孔子的箴言讲述了个人在社会中的处世之学和礼仪之道。孔子在餐桌礼仪方面有很多规矩，因为他认为饮食不仅仅是为了满足个人胃口，也是最重要的社交手段。因此，中国餐桌礼仪看似不足为道，但实际上意义深远。

我经常向我在中国和海外的年轻中国朋友们分享本书。我希望借此鼓励大家通过阅读了解我们中国的传统，了解我们的社会历史，了解餐桌礼仪植于中国文化深处的根基，并为我们的中国文化感到自豪。同样重要的是，我认为本书还可以帮助年轻人在社会中成长，协助他们与来自不同文化的人士交往。因此，我在部分章节前增加了导语，以帮助弥补二十世纪三十年代与今天在人们习惯上的差异。

在所有文化中，共享美餐都与生活中最重要的时刻紧密相关。无论是商务还是休闲用餐，无论是与未来的上司抑或配偶初次会面，适当的餐桌礼仪都将对人们大有益处。

本书已停印多年，市面上很难找到。感谢广西师范大学出版社的支持，本书得以重新出版。也感谢徐忠良先生的支持，徐先生还邀请了曾在纽约联合国总部任翻译的刘炎琦女士将此书部分内容从英文翻成中文。本书保持了原书的中英文格局和插图，希望能帮助读者感受到原书的风貌和魅力。

徐芝韵

纽约，2018年6月

INTRODUCTION

Of all human frailties that of the gourmet is probably the gentlest and the most amiable. His very weakness marks him out as that rare thing, a man of taste and discernment. It can do no harm to anyone but himself and need not harm even him if he has wisdom as well.

To all gourmets I commend this book. If, in these unhappy times, they cannot come to China to see and eat for themselves, this book will build for them an easy and pretty bridge by which, from their own homes, they can reach the simple and surprising delights of Chinese cooking.

In the matter of food, as in all else that has grace, the Chinese have long been past-masters. They put into the cooking of it all the loving care for detail, the daintiness, the spice and the wit that characterize their art and their understanding of the business of life, which is one of the chiefest of their charms. In China every province boasts its own fashion of preparing food and claims that it is the best. When you eat you tend to admit the claim without demur. But a practiced palate will tell in a twinkling whether the juiciness of a duck or the crispness of the crackling of a suckling pig is due to the craft of Cantonese, Pekinese, Szechuanese, or what you will. For myself, I cannot pretend to so much "*expertise*". Nor does this book set out to guide you to it or to cover the vast field of Chinese cooking. It offers 75 straightforward recipes of every day, but none the less delicious, food which is within the reach of everyone, which calls for no expense and which will quench the passions of the greedy and enchant the finer feelings of the gourmet alike.

The authoress, Dolly Chow, is the daughter of Sir Shouson Chow, a well-known and much respected member of the Chinese community of Hong Kong. She has been so kind as to give me some lessons in cooking at which I had the emotion both of finding out the secrets of the Chinese kitchen and of eating the dishes as they came off the fire. If you had seen and tasted what I did, you would agree with me that no one is better fitted than Dolly Chow to introduce you to the dressing of Chinese food.

Maria Teresa Clark Kerr[1]

---

1 Maria Teresa Clark Kerr is wife of British diplomat John Clark Kerr. John Clark Kerr was the British Ambassador to China between 1938 and 1942. Editor's note.

序 一

人类生活之"衣、食、住、行"中最使人觉得不可缺少而使人最觉享受者莫过于食,对于味道之辨别且鲜有人在,人如具有此辨别味道之能,其乐不尽也。

著者以此小书供所有"味食家"之参考。如读者无机会在中国各地尝味者,可参考本书易而得其味也,不但本身可在家中享受,并且可请亲友同尝中国各地之美味。

对于"食",中国有长久之研究及许多得其道者,他们考究各种烹饪,使它成为生活中之一种艺术。于中国各省均有其著名拿手之好菜,其味道各有千秋,难分伯仲,就中以广东、北平、四川等为名菜。著作当不能于此小册中贡献介绍全中国之菜,仅以著名全球之好菜,尤以经济美味为准,选七十五种菜单以供诸位一试。

本著作系一位名厨积几十年之经验所著者,希望诸位不妨一试其味也。

玛丽亚·特蕾莎·克拉克·克尔[1]

---

1 英国外交官克拉克·克尔(John Kerr Clark Kerr)的太太。克拉克·克尔于1938年至1942年任英国驻华大使。——编者注。

PREFACE

The Chinese long ago developed a way of life which has found some favour among visitors to China. The earliest records show that Chinese of the remotest antiquity not only enjoyed the chase; they also had good methods of dealing with the fruits it yielded.

The histories do not state clearly when traditional Chinese cookery methods became fixed, but they do record, in prose and verse, the ecstasies of those men of fine discernment and taste who knew that men ate not merely in order to live. Dr. Johnson's well-known observation that he could always smell a good dinner may readily find its echo from many Chinese scholars. However, only by a long process of empirical trial and error were those subtle flavours discovered which have made men long exiled from home write wistfully of the province which gave rise to a lordly dish.

The author of the book does not attempt to compete with men like Su Tung-po or Yuan Mei, whose writings on the art of cooking have become classics. Her aim is much more modest and her methods are much more practical. She would tell you how, with the minimum of materials and trouble, you may prepare food as the Chinese do; she would demonstrate that it is worthwhile taking a little time to learn the art; and the results are so much more satisfactory. Those who have had the privilege of tasting a meal prepared by her will suspect magic—but the secret is here. If the garden of good Chinese food is an enchanted land (as many have that is) then the author is a good guide.

F.T.CHENG[1]

Chinese Ambassador in London

---

1 F. T. Cheng is a famous Chinese calligrapher and diplomat, who served as the ambassador of the Republic of China to the United Kingdom from 1946 to 1949. He is also a gourmet who specializes in cooking. Editor's note.

序 二

于中国菜，凡访华之外宾无不赞美。在古代，中国不但以擅射猎闻名于世，并且以其食道驰名于全球。历史虽无明确地记载中国烹饪之道从何时开始而广大，但于诗于文，古之雅士确乎记载了味食之喜妙，因其深谙饮食绝非仅为生存之故。有名之约翰逊博士曾说，可嗅之而知一餐之好味道，此言盖与许多中国学者之见者同。但唯有通过长时间之试错，方可得此至味。故一省之著名佳肴常常令流亡之人怀乡而挥毫。

本著虽不及苏东坡或袁枚所著"食谱"之好，但仍可比肩其美。本著注重家庭小饪为主，望常加练习其烹饪之方法。

郑天锡[1]

---

[1] 中国著名法学家、外交家，于1946年至1949年间任中华民国驻英大使，其亦是一位精于烹饪之术的美食家。——编者注。

FOREWORD

Modes of travel have so improved in recent years that the world seems to have really grown so much smaller and international intercourse has become so much easier. Increasing numbers of visitors are coming to China each year and they sooner or later become acquainted with Chinese food. It is natural that many of these people find themselves desirous of learning something of the mysteries of Chinese cooking and the art of imparting the subtle flavourings to the various dishes. There can be no doubt that the taste for Chinese food among foreigners is growing very considerably. The introduction of Chinese food at the well-known hotels in Hong Kong and the increasing popularity of Chinese restaurants in Shanghai testify to the increasing interest of foreigners in Chinese cooking.

Although East and West are getting closer through increased

contact, Chinese customs still differ widely from those of the West and, in some instances, are actually the reverse of Western ideas. To avoid misunderstandings as well as to assist in the full appreciation of Chinese food in the surroundings of Chinese homes and restaurants some knowledge of these customs will be found very helpful:

At a Chinese dinner the host sits opposite the guest of honour because, from the oriental point of view, the highest seat should logically be opposite the lowest. The left-hand side of the host is regarded as being higher than the right-hand side. It is permissible by Chinese custom to remove the short outer jacket, wearing only a long robe for dining. Soup and fish are served only at the end of the dinner instead of at the beginning as is the custom of the West. Warm wine is provided, in contrast to the Western custom of icy cold drinks.

The interior surroundings, in the way of furniture and decorations, are usually very plain in most of the old-style restaurants.

At a Chinese meal such articles as milk, bread, butter or cheese are never served. Sometimes dishes considered by us to be rare and delicious may be thought unpalatable for consumption by foreigners—as, for example, bear paws, chicken feet, duck tongues, pig trotters, etc.

It is not surprising, then, that a stranger confronted for

the first time with unaccustomed articles of our diet should feel dismayed and uncomfortable. As an illustration, I recall the experience of a friend of mine, an English lady of title, who was on her first visit to this country to see her son, the head of a large foreign tobacco company in Peiping. Her fist encounter with Chinese food was at a dinner in a restaurant. The unaccustomed dishes upset her so badly that she practically decided not to accept any further invitations to Chinese dinners. This was the excuse she made to me when I asked her later to dinner. However, she finally persuaded herself to try once again, and accepted my invitation. Her second adventure was more fortunate, and she told me that on this occasion she found the food really delicious and was now anxious to make a third trial. From that time onwards my friend would eat and enjoy herself even in restaurants. She returned to England many years ago. In one of her letters to me she reminds me of the delicious dishes she had tasted while in China. Had this lady not remained long enough to have had the opportunity of discovering really attractive dishes, she most likely would have concluded that Chinese food was revolting, and would have conveyed this impression to the people back in her own country.

Many of my foreign friends have expressed the desire to see, in a handy form, an explanation of Chinese customs and practices supplementing the information on the preparation and enjoyment of Chinese food, and have thus given me encouragement and confidence to attempt the task in this little book. I also desire to assist my friends to prepare a few dishes in their own homes so that they may be able to demonstrate how delicious Chinese food really is.

It is my sincere hope that this little volume will be helpful to all epicures who desire pleasure and enjoyment from the exotic as well as the delicious in food.

D.C.

自 序

近年来,旅游方式有了很大的改善,世界似乎变得如此之小,国际交流也变得更加容易。每年越来越多的游客来到中国,他们或迟或早了解到了中国的饮食。自然有很多人希望学习中国烹饪的神秘之术和各种菜肴调味品的微妙艺术。毫无疑问,外国人对中国食物的喜好正在迅速增长。香港著名饭店纷纷推出中餐菜式,中式餐馆在上海人气急升,这都证明了外国人对中国烹饪的兴趣与日俱增。

尽管东西方之间的交流越来越密切,但中国的风俗习惯仍然与西方国家大相径庭,在某些情况下,实际上与其恰恰相反。为了避免误解,同时协助外国人在中国的家庭和餐馆的环境中充分品赏中国饮食,这些关于习俗的知识对他们将非常有帮助:

在中国的宴会上,在主人对面落座的是最尊贵的客人,因为从东方的角度而言,最尊贵的席位循理应放在地位最低微者的座位的对面。主人左手边的座位被视为地位高于右手边的座位。中国习俗允许人脱掉短外套,只穿一件长袍进餐。汤和鱼只在宴会临近尾声时食用,而不像西方那样在宴会开始时食用。与西方传统的冰镇饮料相反,中国宴会提供的是温酒。

大多数老式中餐馆的室内环境,就家具和陈设而言,通常是非常朴实无华的。

在中餐中，从不提供诸如牛奶、面包、黄油或奶酪之类的食物。有时，我们认为难得的美味可能不会被外国人接受，例如熊掌、鸡脚、鸭舌、猪蹄，等等。

一个外国人面对我们的饮食第一时间可能会感到沮丧和不安，这并不奇怪。举个例子，我记得我的一位朋友——一位有身份的英国女士——第一次到中国来看她的儿子，她儿子是北平一家大型外资烟草公司的领导。她与中国饮食的第一次相遇是在一家餐馆，难以习惯的食物让她决定不再接受任何中式宴会的邀请，这是我后来请她吃饭时她推辞的理由。然而，她终于说服自己再试一次，接受了我的邀请。她的第二次冒险幸运得多，她告诉我，这一次她发现食物真的很美味，现在她急于要做第三次尝试。从那时起，我的这位朋友甚至可以自己在餐馆享受中国美食了。许多年前她回到了英国。在写给我的一封信中，她提及了她在中国尝到的那些美味。如果这位女士在中国待的时间不够长，她就没有机会发现真正有吸引力的美味，她很可能会得出结论：中式饮食是令人反感的，并且可能在回国后将这个印象传递给其他人。

我的许多外国朋友都表示，希望看到有谁能用一种便利的形式，解释中国的习俗与中式饮食的制作和食用方法。这给了我鼓励和信心，让我尝试用这本小书来完成这个任务。

我也希望能帮助我的朋友在他们自己家里准备一些中式饮食，这样他们就可以向他人展示中式饮食是多么美味。

我真诚地希望这本小书将有助于美食家们从食物的美味与异国情调中获得愉悦和享受。

周德丽

Plain Daily Meal
日常饮食

THE HOUSEWIFE AND COOKING

Even though the anachronistic term "housewife" has properly fallen out of favour, and that hosts and hostesses now come in all shapes, orientations and persuasions, the principles herein still apply, maybe even more so than in Dolly's time.

Foodies, farm-to-table enthusiasts and discerning consumers of regional cuisine will embrace the Confucian principles of fine dining. No well-heeled fan of fine Chinese cuisine can match Confucius's fastidious approach to food.

For example, his refusal to eat dried meat and drink wine bought at markets reflects his preference for the freshest home-grown goods possible. Today, Confucius might do his shopping at farmer's markets, favouring organic produce and free-range artisanal meats, and the offerings from micro-breweries and small vineyards. We would do well to follow his example.

His preference for eating locally-grown foods in season now seems like common sense, while his liking of finely minced meat and polished white rice reflects refinement and an aesthetic concern for presentation—all of which speaks to a hostess/host's concern for her or his guests.

Carolyn Hsu

One of the duties of a good housewife is to appreciate the importance of cooking and its close relation to domestic happiness. This opinion prevailed even in the days of Confucius. We learn from the classics that Confucius was particularly fastidious as regards food for we are told that:

1. He refused to partake of wine and dried meat bought in the market.

2. He refused to eat meat which was not cut properly nor what was served without the correct sauce.

3. He refused to eat anything badly cooked or not in season.

4. He liked his rice polished white and his meat minced finely.

Had there been such conveniences as are now available—the equipment of a modern kitchen—and had there been properly conducted schools of cookery where the necessary knowledge might be acquired, Chinese housewives in ancient times would have been able to maintain satisfactorily a high standard of culinary art in their households. The modern housewife is more fortunately situated. She can employ a good cook and supervise his work, thus doing away with the necessity of purchasing food from outside herself. She can procure a mincing machine to mince meat to any desired fineness; she can buy perfectly white machine-milled rice and she can easily procure any sauce to suit any taste thanks to the skill and experience of modern organisations. In fact she has everything in her favour these days; all she has to do is to

The Housewife and Cooking 家庭主妇与烹饪

learn how to cook.

One would expect then that there should be less friction in modern households. Unfortunately this is not so. Perhaps it is that Confucius, being wise, was exacting only in the matter of food, while modern husbands are fastidious and troublesome in other directions as well. However, experience tells us that, by eliminating one cause of domestic disturbance—the food question—the modern housewife has gone far towards securing domestic peace. "Good appetite brings happiness" (食得是福) is an old saying worthy of the housewife's attention.

To stimulate the appetite is the one object of our culinary art, the knowledge of which enables the housewife to produce dishes so deliciously flavoured and so attractively served, that they would tempt even the most fastidious husband. The same knowledge will also help her to bring the changes in the diet which, like a change of air, can only be beneficial to the appetite and health.

To the conscientious housewife then, who is solicitous of domestic peace and happiness, the science and art of cooking should have a definite appeal. The servant problem fortunately is not so acute in this country; so the housewife, once she has acquired proficiency in the art of cooking, needs to have less uneasiness of mind in regard to the menial work of the kitchen. Her part will merely consist of direction and supervision, if necessary. The Chinese say, with truth, that just as those who live near water know the nature of fishes, and those near mountains learn the melody of birds, so those who remain close to the kitchen acquire the knowledge of good food (近水知鱼性，近山识鸟音，近厨得佳食。).

GOOD APPETITE BRINGS HAPPINESS

食得是福

THE ART OF COOKING

China is a country where the appreciation of good food is developed into a fine art. Chinese are epicures. Their cooking is distinctive: No other cooking resembles it in any way.

Chinese food is rich, but not greasy; it is delicately flavoured, but not pungently spicy. Cook what is freshly slaughtered, and eat what is freshly cooked (现杀、现烹、现熟、现食) is a doctrine universally recognised throughout China. It is better that one should wait for the meal than that the meal should wait for one. Variety is another important feature. A Chinese dish almost always consists of a mixture of foodstuffs—the meat or fish is generally cooked with, and improved by, the addition of some appropriate vegetable. All the material to be used is cut into convenient size in the kitchen before serving, so that no carving instruments are required at the table. All the condiments are added during the process of cooking, thus doing away with the necessity of the usual cruet. The only exceptions are some soya bean sauce and vinegar provided at the table in case they are required.

With the passage of time the methods of cooking have necessarily undergone many improvements as compared with the original crude processes. Expert cooks in different parts of China have introduced numerous improvements, and, with China being such a vast country, its component parts differing widely not only in climate and customs, but even in the spoken language, it is only to be expected that different terms are found in different localities for the same way of cooking. For instance, roasting in the North is known as *K'ao* (烤) while in the South it is called *Shao* (烧). Similarly *Shao Fan* (烧饭) in North China means cooking rice, but in Canton they say *Chu Fan* (煮饭). In these circumstances I have to employ those terms which are more commonly used and are more generally understood. All the terms used in this little book are in the National language, that is, Mandarin [*Kuo Yü*(国语)].

METHODS OF COOKING

1. Shao K'ao (烧烤) *means Roasting.*

There are two different ways of doing this: one is roasting over an open fire known as *Ming Lu* (明炉), while the other is roasting in an oven *K'ao Lu* (烤炉).

By the first mentioned method we prepare roast suckling pig and barbecued Peking duck. In exactly the same way the Russians make "shaslick"[1] and the Javanese "sateh"[2] dishes, which are well known to foreigners in the Orient. The Cantonese dish known as "gold coin chicken" consisting of a combination of alternate pieces of ham, chicken and pork is made the same way. Material for barbecuing should be hung for 6 to 7 hours, and then covered with the proper condiments. Then it is fixed to a metal fork or skewer and held over a strong charcoal fire. Constant turning of the forks is necessary to ensure even roasting. In barbecuing a whole pig the skin should be punctured

---

1 It should be "shashlik", one kind of barbecue, which was originally popular in eastern Europe, the Middle East, parts of Central Asia, and spread to Moscow at the end of the 19th century, and quickly became popular. By the 1920s, it was already a street snack in all Russian cities. Editor's note.

2 A Southeast Asian skewer, also known as "satay", popular in Indonesia, Malaysia, Thailand, Singapore, etc., was invented by Javanese street vendors. Serve with a variety of sauces. Editor's note.

The Art of Cooking   烹饪的艺术   9

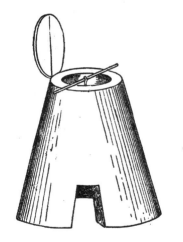

A Chinese Oven
中国炉

before roasting to secure an even surface at the end of the operation.

A Chinese oven is usually built of brick and clay, with two openings, one below and one above in the form of a very short chimney. A charcoal fire is started inside. After half an hour when the oven is sufficiently hot, the fire is damped down by placing a thin metal sheet over the charcoal. The material to be roasted is hung inside the oven, and both openings are closed. The result is identical with that of a modern gas or electric oven.

2. *Chêng* (蒸) *means Steaming.*

This is a simple and economical process of cooking with steam rising from boiling water. By this method the nutritive juice and flavour of the material are conserved. It is specially recommended for preparing fish. In the home, steaming is very often carried out by utilising the steam from boiling rice. At other times it is done by means of a special cooker made of split bamboo called *Chêng Lung* (蒸笼) meaning a steam cage. In steaming *Chiao Tzu* (饺子) or dumplings, these are placed directly on a piece of damp cloth spread over the cage instead of on China plates. This method of cooking is frequently used at home, but not often in restaurants.

3. *Ch'ao (炒) means Frying with a little fat over a quick fire. Frequent quick turning is essential.*

Most of the expensive dishes are prepared by this way of cooking. The important point in this process is the high temperature employed, which ensures quick cooking. If the temperature is not sufficiently high, the meat will not be tender. On this account it is not advisable to fry more than a pound of meat at a time: if more is needed the process should be repeated several times. When more than one ingredient has to be dealt with, they should be first fried separately and then mixed together, the meat being left to the last always. The Chinese frying pan being convex, and not flat bottomed, the turning is more conveniently done.

4. *Chien (煎) means Sautéing or Frying with a small quantity of fat over a gentle fire. Turning is necessary only when the meat has sufficiently browned.*

This process is also applied as a preliminary treatment to poultry and meat before stewing or boiling, for two reasons. Firstly, it makes the meat more palatable. Secondly, it tends to eliminate any excessive "muttony" or "fishy" flavour.

5. *Cha (炸) means Frying in deep fat at a high temperature.*

The substance to be cooked by this process is immersed

in boiling oil. Sometimes the meat is covered with a coating of flour before it is dropped into the oil, to prevent toughening or overcooking. Meat treated by this process is very indigestable, though it is very tasty, so it is not advisable to employ this method too frequently at home.

6. *Mên* (焖) *means Stewing.*

By this method the meat and vegetables are cooked together with a small quantity of water at a moderate temperature for a long time. The juices of the meat and vegetables are retained in the liquid and the long and slow process of cooking renders the material both tender and delicious.

7. *Tun* (炖) *means cooking by use of a double boiler.*

The material for cooking is contained in a covered vessel like a casserole-dish placed in a pan of boiling water. Meat so treated is more palatable than by simply boiling. More water is required than in stewing, otherwise the process is much the same. This is the best way of making soup, when special care should be exercised to seal down the cover with wet tissue paper. In winter this process is very popular.

8. *Ao* (熬) *means Simmering.*

This is a simple process of boiling very slowly. Care must

be taken to avoid too rapid evaporating, and not to uncover the pot too often. The simmering process must be continued without interruption until the time of serving. It is also termed *Po* (煲).

FLAVOURING

Various flavouring agents are employed, the most common being onions (葱), ginger (姜) and garlic (蒜头). Reference to the ancient writings seems to show that they are made to serve a double purpose, for, apart from imparting a distinct flavour to the food, some of them are believed to have medicinal qualities as well. For instance, the onion, which forms a constant ingredient in the preparation of fish, is supposed to be able to counteract any possible fish-poisoning, while ginger is reputed to have stimulating properties. Garlic and bean relish (豆豉) form an excellent flavouring for fish or pork. Red pepper [*Hua Chiao* (花椒)] and star aniseed [*Pa Chiao* (八角)] go well with beef and duck. The usual method employed for the addition of flavouring is as follows:

Heat a little lard in a frying pan, then add the flavouring agent, only a small quantity being used—a piece of Chinese onion, a slice of ginger or a clove of garlic (sliced). Fry until a light brown. The material to be cooked is now added and the cooking is continued as planned.

USE OF FLOUR

The use of flour when shaping meatballs or thickening gravy is termed *Ch'ien* (芡). Sometimes a coating of flour is added to meat destined for frying to prevent it from being too easily overdone. Paper has been known to replace flour for the same purpose. Bean flour (豆粉) or caltrop flour (菱粉) are commonly employed by Chinese. Cornflour (corn starch) is equally good for this purpose.

SELECTION OF INGREDIENTS

The quality of the ingredients plays an important part in good cooking. Just as no good painter can work well with a broken brush or dried up paints, so no good cook can succeed without the correct kind of ingredients. For instance, in choosing chickens, poulets are more desirable because their meat is more tender; on the other hand, with duck the male species is preferable. For stewing, steaming and the Chinese way of frying, use spring chickens; for roasting choose capons, while for making chicken soup, when the meat will not be served, old birds may be used. Fillet should be used for frying, flank for making meatballs, and loin for roasting. Eat everything in season, and you will get more palatable and more economical dishes. *Sam Lai* (三黎) or *Shih Yü*

(鰣鱼) (i.e. shad) is, as its name implies, a seasonal fish, and when in season is in great demand. The fish swims up the Yangtsze and other rivers to spawn. On the return journey, its eyes become reddish in colour and it is not as fat as when it started. For this reason red-eyed *Sam Lai* is considered inferior in quality and can be obtained at cheaper prices.

In the case of vegetables the "heart" only should be used for cooking. Bamboo shoots and mushrooms can be prepared with or without meat. They are equally delicious either way.

SERVING

The appearance of the food when served is another important factor which should not be overlooked. Well-arranged dishes attract the eye and when a pleasant flavour accompanies them, the appetite cannot but be stimulated. This is a general rule; it applies to all forms of cooking, Chinese and foreign alike. Vegetarians fully recognise its importance: they have giblets, roast duck, boiled chicken and other dishes, all of them being prepared with bean curd "skins". The dishes are so cleverly made that they look exactly like the real article. Colour schemes are also made use of to decorate food, particularly the sweets. A good chef will not be satisfied to produce dishes which tickle only the palate; they should also gladden the eyes, and their names should please the ears as well. The best example is a Cantonese dish called *Pi Yü Shan Hu* (碧玉珊瑚) which means "green jade and red coral". The green jade is represented by vegetable stems, and the red coral by the fat of crabs. This dish is beautiful in appearance, delicious in taste, and in addition, elegant in name.

DINNER PARTIES

Today, as in Dolly's time, we still host dinners at home, at restaurants and at special venues like banquet halls. Some are sit-down affairs while others are more informal buffets. The menu format is largely unchanged, but more options are now offered to accommodate various regimens—pescatarian, vegetarian, vegan, gluten-free, low-salt, low-fat. For sit-down dinners, it is quite acceptable to ask guests in advance if they have dietary restrictions. For buffets, it is simply a matter of offering a wider selection of dishes. In general, it is usually a good rule of thumb to avoid very salty or heavy sauces on your dishes—with a salt shaker close at hand, guests can salt their plates to their heart's content.

Special attention should be paid to the guest-list—like your favourite dish, the perfect dinner party is an adept blend of ingredients.

As in Dolly's day, if the host is not a talented cook, a caterer or chef will be hired for the occasion. Chinese protocol has it that if the dishes warrant it, the host will laud the guest chef with high praise, while reserving self-effacing criticism for the host's own cooking, no matter how superb. This is our Chinese culture—to be humble, no matter what circumstances fortune has bestowed on us.

Carolyn Hsu

RESTAURANT DINNERS (酒席)

A dinner at a restaurant can be ordered in the form of a table d'hôte (整桌) or à la carte (小酌). The latter is only for dinners of a very informal nature among intimate friends, and must never be given in entertaining an honoured guest or in celebrating some important event.

A full course dinner usually consists of eight large and eight small dishes (八大八小) with the addition of pastries, rice or noodles and fruit. It is a long menu, and is, in truth, a great deal more than necessary. Ten years ago, however, such a menu would have been considered surprisingly short, as, at that time, a full table often consisted of about forty dishes or courses.

The dinner begins with four cold dishes which are placed on the table before the guests take their seats. Cold ham, an important item, is always placed before the guest of honour. Occasionally sliced duck takes the place of ham. In recent years there has been a tendency to combine these four small dishes into one large dish called *P'ing P'ên* (拼盆). Following these, four hot fried dishes or *Ch'ao Ts'ai* (炒菜) will be served one at a time. These always consist of something in season.[1]

---

1   Nowadays, tables in even the most high-end Chinese restaurants are equipped with a revolving platform or "lazy Susan". Dishes are placed on this platform which is then turned (sometimes by the obliging host) so that guests can serve themselves as each dish presents itself in front of them. Editor's note.

Then comes the main part of the dinner, the eight large dishes. According to Chinese custom the best should be served first. Shark fins, being considered the foremost delicacy, therefore take the lead. In North China bird's nest ranks equally as high. A couple of other dishes are next served. Then comes the second important dish which is usually a roast, such as barbecued duck or suckling pig. The rest follow in turn, a fish and a soup always making the last items. Chicken soup is a favourite in the South, while the Northerners prefer duck soup. The meal concludes with rice served in small bowls, dainty pastries and a large bowl of some sweet liquid like hot orangeade or almond tea.

HOME DINNERS (家厨酒席)

A higher standard of cooking is required in the preparation of a home dinner. Every detail of the culinary art can be more conveniently carried out in one's own kitchen. As it is more important to aim at quality rather than quality, it is usual to provide a dinner of only eight or ten dishes, which should include a choice soup. Some foreigners, perhaps mistakenly or jokingly, call a home dinner "coolie chow", but, as a matter of fact, the most pleasing and tasty dishes are often met with at these meals.

Since the inauguration of the New Life Movement it has become a recognised practice to entertain at home, the food being prepared on the premises. If the host does not possess a cook sufficiently skillful for this purpose, he usually gets around the difficulty by enlisting the services of some capable cook known to him. It often happens that the cook thus secured is in the employ of one of the host's many friends: he borrows the cook, and at the same time invites the employer to the dinner. At the end of the repast the host will remark on the excellence of the food served, while the friend, whose cook's services have been requisitioned, will say just the reverse. The other guests, if ignorant of the arrangement, are naturally astonished at the lack of modesty on the part of their host, and the seeming rudeness on that of the friend. When the secret is disclosed, it is only right that the host should praise the accomplishments of his friend's cook, while the friend should remain modest in regard to them.

A dinner at home has many advantages over one at a restaurant. The guests can be made more comfortable, and are permitted greater leisure in the enjoyment of the repast. Cleanliness is more likely to be observed in the preparation of the food, thus making it more wholesome. Lastly, it is generally more economical.

When a regular dinner is given at home, the host, from modesty, always calls it "plain dinner" *Pien Fan*（便饭）.

PLAIN DAILY MEALS

The daily meal is called *Chia Ch'ang Pien Fan*（家常便饭），

i. e., ordinary plain home meal. For a family of five or six persons a daily meal usually consists of three meat dishes and two vegetables. Soup may or may not be served. The main point of difference is that during the daily meal, all the dishes and the rice are served at the same time. Chinese are taught from childhood to regard rice as the main item of a meal and to partake only sparingly of meat and vegetables. Though there may be plenty of meat and vegetables, children are always taught not to take too much of it.

家庭主妇与烹饪

尽管"家庭主妇"这个过时的词已逐渐失宠，如今的男女主人们已各持有自己的新姿态、新取向、新主张，但此处提及的饮食之道仍然适用，或许甚至比周德丽的时代更为讲究。

美食家、有机餐饮爱好者以及对原产地美食情有独钟的人都会欣然赞同孔子对于美食的原则："食不厌精，脍不厌细。"恐怕没有任何中国美食的拥趸可以媲美孔子对食物的挑剔。

例如，孔子不食用市场上卖的干肉和酒，他喜欢尽可能新鲜的本土食材。如果生在今天，孔子可能会在农夫市场买菜，偏爱有机蔬菜、散养肉禽、小型手工啤酒厂和小型酒庄的精酿。如果遵循孔子之道，我们会生活得很好。

孔子喜食本地当季食材的偏好现在似乎已成为常识，他还喜欢肉切得薄些细些，米饭要颗粒完整、不带谷壳，这都反映了对食物精致、美观的要求——所有这些都代表着主人对宾客的心。

徐芝韵

一个主妇之重要责任系知道如何烹饪,并且能烹饪一手好菜,这关系家庭之伦乐,无论古今中外均是如此。吾人由于古典可知孔子曾亦说过关于食之问题,吾人能知:

一、他拒酒与干肉购于市场者。

二、他拒食物无适切者及无适饪者。

三、他拒食物无合季者及不善饪者。

四、他喜欢白饭及细切之肉。

古代哪有现代可具配之厨房?哪有如现代之好技术及烹饪之学校?古代的主妇们仅能由她们的家庭学习烹饪之方法,不如现代之主妇们之幸运。现代主妇能享受一切古代所未有之好设备及教材,不仅在家亦可在学校学习到,更有如本书介绍各种各样各地各省之好菜者。当然她可雇用厨师替她烹饪好菜,她可用现代优良之烹具以供烹调,她也可向菜馆叫菜,但一个主妇最重要者莫过于学习如何烹饪一手好菜以维持家伦之乐。

许多人可能认为一个现代之主妇当可马虎点了,其实不然。一个家庭之乐处均靠主妇之烹饪,假如现代的主妇们忽略了食饪,那么亦可就说她忽略了家庭之伦乐矣!古语曰"食得是福",现代之主妇们当应注意此事也。

刺激食欲是厨房艺术之一种目标。此种艺术能促使主妇做很美味的菜、很诱食欲的菜,这能使你那严格庄重的丈夫高兴。此种艺术将帮忙主妇变化菜单,如令空气焕然一新,使家庭里的人健康快乐。

对于一个贤妇，家庭的和平及幸福是不可缺少的，而这种烹饪的学问及艺术确实是一个很好的帮助。幸而仆役的问题在本国是无甚尖锐严峻的，因此一位贤妇如能得熟了烹饪的艺术，就可不必注重在厨房粗杂琐碎的工作而易于造出好菜了。她如需要可仅于指导而监督的位置。有一句古言说得很真实：近水知鱼性，近山识鸟音，近厨得佳食。

烹饪的艺术

中国是一个重食的艺术的国家,并且关于食道是非常考求的,因此中国人多半是食通。他们的烹饪是不同的,世界上没有其他菜具与中国菜相似美味的。

中国菜是很丰润而不太油的,那是富于美香风味的,而色、香、味俱全。中国菜之现杀、现烹、现熟、现食是举国之共识。人们等菜烹熟比烹好菜来等人们食强得多了。中国菜花样之多是一个很重要的特点。中国菜通常用多种的原料混合做成——一道菜里通常有鱼或肉,并且附带了一些可口的蔬菜。所需烹饪之材料先切成之后才开始烹饪,所以不必准备小刀在餐桌上。所需之香料调味且在烹饪时加入之,因此亦不必在餐桌上排齐之盐和胡椒粉等调味料,仅要为酱油和醋而已。中国菜的烹饪必定有很多复杂之调味品,在烹饪时加入之,西餐是自己加以调味的。

中菜的厨师不断地在改良其烹饪之法,而使各样之菜成为更可口。中国的地方是那么广大,不但其菜的烹调各有不同,为了气候风俗之不同其风味各有千秋之感,并且语言之复杂,同一烹饪法之称呼亦有不同。就举例说,"roasting"在中国北方称为"烤",而南方且称为"烧"。同样,北方称为"烧饭",而南方且称为"煮饭"。在这种情形之下,著作以通常之称法加以说明,为使读者们知之,所有称呼均以"国语"注明之。

烹饪之方法

1. 烧烤

有二种不同之方法:一种系明炉烧之,另一种是密盖之炉(烤炉)烧烤的。

第一种所说的方法常用于制作烤乳猪及北平烤鸭。此种方式完全与世界知名之俄人之"Shaslick"[1]或爪哇人的"Sateh"[2]菜一样。吾人所知之著名广东菜金钱鸡(Gold Coin Chicken),内含有火腿之切片及鸡肉猪肉,同样以此方法造成的。要熏烤之东西必挂钩着六至七小时,然后加上适当的调味料,然后以金属叉串着或以曲钩挂起以强火烧烤,并且不断地将叉转动使之全部烤熟。整个小猪在烤之前在皮上以小针细打小穴,加入之调味料,但不可损害全面皮肤。

中国的烤炉是砖泥所造成的,有两个孔,一在上面一在

---

1 应为"Shashlik"(下文同),一种烤肉串,最初流行于东欧、中东、中亚部分地区,十九世纪末传至莫斯科,迅速风靡。到二十世纪二十年代已经是遍布俄罗斯各城市的街头小吃。——编者注。
2 一种东南亚烤串,又称"satay",流行于印度尼西亚、马来西亚、泰国、新加坡等地,由爪哇街头小贩发明。吃时配以各类酱汁佐餐。——编者注。

下面，好像一个小烟筒。以木炭在炉内烧之，起火约半小时后，炉内之热度已够热，用薄金属板放置在火上。要烤的东西挂在炉里面，并且将两孔均塞住。效果与现代之瓦斯炉或电气炉一样。

2. 蒸

　　这个用沸开了的水之蒸气来做食物的方法是经济的烹饪法。以此方法可保存有营养的汤汁及风味，特别是可保存新鲜之鱼味。在家中吾人常常利用烧饭时之蒸气。此外吾人可用蒸笼来烹饪之。用蒸气做的有饺子和馒头。这种方法通常是家常之饪法，在食堂餐馆是较少采用的。

3. 炒

　　炒，是以少量的作料在热锅里以快动作搅炒的。

　　大部分高贵的菜以此法所造成。最要紧的是锅子要热，以确保快速炒成。如热度不够，则食物会失去应有的味道。因此要炒的材料一次以不超过一磅[1]为原则，如系需要就要分数次炒之。如炒的东西不仅一种，就分次炒完，混拌之，肉则留于最后炒之。中国之炒锅并非平底，故易于混炒之用。

1　1磅约为0.453千克。本书为保留原貌不改。——编者注。

4. 煎

以少量的油用文火烹之曰煎。直到肉熟至金黄时，方有必要翻面。

肉或鸡鸭在焖煮之前也可用此法做预先处理，原因有二。其一，此文火煎则可保存其原味及新鲜。其二，可除其腥膻之味。

5. 炸

以高温油炸。如要炸的东西先准备好，浸入开沸了的油锅中。有不注意肉会过老或过熟焦黑的，以避免起见，筛上面粉后炸之较佳。此种东西虽然很好吃，但在消化方面较慢，在家庭中此种菜是很少采纳的。

6. 焖

此种菜是以蔬菜及肉以少量的水沸煮的，以中火长时间沸煮之，至菜肉均软烂，汤及菜肉均很可口。

7. 炖

此种方法是外锅放水，内锅放要做的菜料，以中火长久烧之，以沸水蒸气炖熟菜。外锅之水当要多一点。用此种做法做汤最佳，此时锅盖须以纸巾封严。在冬天此种菜是很普遍的。

8. 熬

把要做的几样菜放入锅内与水共煮之,文火长时间,忌频繁去盖,忌大火沸煮之。亦称煲。

调味法

中国菜中调味料种类很多,其主要者有葱、姜及蒜头。以古书为参考则可知它们有两种用处,一为食物之调味,一为药料之用。举例说,葱作为烹鱼的常用调味料,被认为可中和鱼中可能存在之毒素,而姜则可当为激剂而齐名,蒜头及豆豉是鱼、肉之调味不可缺乏之圣品,花椒、八角则为牛肉及鸭子之调味用品。通常其调味用法如下:

肥肉切一小块放入热锅先加热之,后放入调味料如葱、蒜头(需切为小片)或姜丝。炸到略见褐色,则可放入所需煮炸之菜料。

面粉之用法

做肉丸或使肉汤增稠的面粉称为芡。有时为了炸鱼、肉怕过度而以面粉被之以炸。纸张有时可以代替面粉包裹炸物。在中国,豆粉或菱粉常代替面粉,玉米粉亦可以代之。

选　料

　　选料是烹饪之一个重要的因素，好比一个书法家不能以一支坏笔来写好字一样，所以一位好厨师不能不选好材料。比如说选鸡必选其肉细软者，另一方面鸭子则必选雄的好。在中国菜中焖、蒸、炒常用春鸡，烤烧则常用雄鸡，如做鸡汤而肉则不取用，那可用老一点的鸡为佳。里脊肉通常油炸之，胁腹肉通常作为肉圆，排骨肉则烤烧之。

　　吃当季的东西，你会得到更好的、更实惠的菜肴。鱼类亦如此。三黎或称鲥鱼是种季节鱼，当季节到来时，该鱼由长江而上，此时肥而可味；季节过时，该鱼则顺流而下，此时不但瘦而眼睛变为红色，因此红眼三黎被认为是次品，所以价钱便宜。

　　蔬菜只取其"心"用于烹饪。竹笋及香菇既可与肉同煮，亦可单独烹饪，同样地好吃。

侍　餐

　　做菜是一种要学之事，但亦不可忽视侍餐之规。精心摆盘布置之菜肴可愉悦双目，同时如伴有香味扑鼻，则更可令胃口大开。这个规则适用于各种菜，中国菜与外国菜一样。素食者充分认识到它的重要性：他们用豆腐皮做出内脏、明炉烧鸭、白斩鸡等菜肴,这些菜如此巧妙,看起来像真的一样。做菜的色彩艺术像艺术家绘画上色同样重要,如颜色的配合别有美的感觉。一个好厨师不会满足于会做可口之菜,亦必善于安排配合,不但美观而其名称亦必善耳。举例来说,广东菜则有"碧玉珊瑚",此种菜不但好看而味美,其名亦非常配合动听。

宴 席

今天,和太舅婆的时代一样,我们仍在家、餐馆以及宴会厅等专门场所设宴款待宾客。有些是正式宴会,有些是非正式的自助餐。菜单形式基本没有变化,但现在提供的选择更加多样,以满足不同人群的饮食习惯——海鲜素食主义者、素食主义者、纯素食主义者、无麸质饮食者、低盐饮食者、低脂饮食者,等等。对参加正式宴会的宾客,可以提前询问他们是否有任何饮食禁忌。至于自助餐,提供更多菜肴选择即可。一般而言,一个很好的经验法则是避免提供过咸或酱汁浓重的菜肴——在餐桌上提供盐瓶,客人可以自行取用。

宾客名单应特别注意,和一道美味佳肴需要相互搭配的原料一样,完美的晚宴也需要相得益彰的宾客。

在太舅婆的时代,如果主人不善烹饪,会请厨师来帮厨。中国人的礼节是,如果菜肴受到欢迎,主人会对帮厨的厨师大加赞赏,同时自谦厨艺不佳,无论实际上主人的厨艺多么高超。这就是我们中国文化——在任何情况下都要谦虚。

徐芝韵

酒 席

在食堂（饭店）之酒席可订整桌（table d'hôte）或小酌（la carte），后者为很亲之朋友们聚餐之用而不必用请帖招待的，绝不能招待贵宾或用于庆祝一些重要事。

整桌是通常有八大八小的菜，加饭或面或馒头，以及水果。菜单是很长的，实在说，那是太奢华了一点。然而十年前，这种菜单仍被称为太少了呢，在那时候，整桌之菜是约四十种菜呢！

酒席以四盘冷盘开始，在客人就位前需先排好在桌上。一个最要紧的是火腿，通常排在最高位客人之前，有时亦有以鸭肉片代替火腿。最近几年则有四小冷盘合为一大盘之趋向，称为拼盘。跟着就有四个热的炒菜，一次上一个，这些菜是以季节而异之。[1]

然后酒席之主要的八大菜就随之而来。以中国的风俗习惯，好菜是先用之。鱼翅被认为是最高贵的名菜，故先出来。在中国北方，燕窝与鱼翅是同样地高贵。随后，有一对其他的菜配之。然后次要之菜就出来了，通常是烤鸭或烤乳猪。然后跟着几种菜，鱼及汤是排在最后。鸡汤是南方著名，但北方是喜欢鸭汤。饭是以小碗，面是以大碗装之，最后有一大碗甜汤，好比热橙汁或杏仁茶。

---

[1] 今天，即使是最高端的中餐厅也会在餐桌上装置圆转盘。菜肴被置于这一可转动的平台之上，转至宾客面前（有时这一任务由殷勤的主人来承担），供客人们自由取食。——编者注。

家厨酒席

准备家厨酒席,高水准之烹饪是最重要的。每一样菜都可充分表显家厨之高明及艺术,那是重质而不重量的,平常是八至十道菜,包括一道精选的汤。有一些外国人可能是错误或开玩笑地称家厨酒席为"coolie chow"(苦力餐),但事实上来说,最愉快而最可口的菜通常在这种席上才能享受到。

自从"新生活运动"开始以来,在家待客成为现代家庭之重要部分,食物之考究亦随之重要。假如主妇们无烹饪之上好技术,则常常遇到备餐之困扰。许多人因此而雇厨师,或在请客时临时雇用所识之良厨。以此常有人借雇友人之厨师,并请友人列席宴会。餐毕,宴会主人赞佳肴之美而此友人则相反之,令其他宾客惊于主人之不谦与客人之不敬。然一旦明其就里,则知其顺理成章。

家厨酒席比之餐馆,在请客时不但宜谈笑风生,各位来宾自谈更有羡慕主人食福,以此可多交友人,食物亦健康卫生,非常经济合算。

正餐在家中所做通常谦称为"便饭"。

家常便饭

家常便饭,就是家中通常之饭菜。五六人之家庭中,便饭通常有三荤两素菜,汤可有可无。主要不同者在家中之便饭是菜饭均同时食用。中国人从小就被教导认为米饭系主要食物,故即便可能有许多菜,孩子们还是常被教导不要贪食之。

A Home Dinner
家厨酒席

TABLE MANNERS

The importance of table manners cannot be overstated. They are the ground rules that allow people from different social and cultural backgrounds to interact over a satisfying meal. It is as true today as it was 2,000 years ago in Confucius's *Book of Etiquette*. Good manners are a way of showing your fellow guests respect and conviviality, of organizing what could be a chaotic event into a seamless, joyous and comforting experience. Given how global our society has become, with international travel and foreign job postings more common, and with our table being filled with guests from every cultural tradition and every rung of society, these ground rules are even more essential. Good manners are an international language that allows one to feel at home at any table.

Carolyn Hsu

SEATING ARRANGEMENTS

The Analects state, "He who exercises government by means of his virtue may be compared to the North Polar Star, which retains its position, and all other stars always turn towards it." The Emperor was known as the Son of Heaven, and his throne was always placed on the North side of the Hall facing the South. The highest seat is therefore situated in the North facing the South. Consequently, in arranging seats we have to take the following into consideration:

(a) A *K'ang* (炕) may be considered as a layman's throne. A "long table" is a narrow high table on which offerings to God or to ancestors are placed. When there is either one of these two things in the room, the side on which it is placed is taken for granted as the North, regardless of its true direction.

(b) When both are absent from the room, the entrance is always regarded as the South.

At a round dinner table the seat of the guest of honour is on the North and that of the host opposite to him on the South. The left side is considered higher than the right. The reason is not known, but it is presumably due to the fact that the sun rises in the East which is on the left-hand side looking from North to South. To the left side of the guest of honour then, is the next seat in rank and to his right the third, and so on. Those

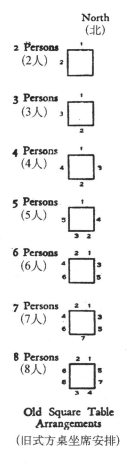

Old Square Table Arrangements
(旧式方桌坐席安排)

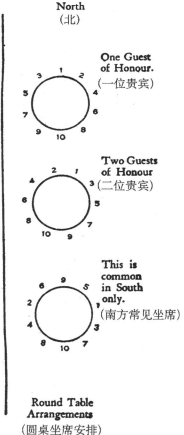

Round Table Arrangements
(圆桌坐席安排)

who sit near the host are either minor guests or his very intimate friends.

In South China the seating arrangement amongst the merchants is different, the seat directly opposite the host being considered the next above the host himself. The guest of honour is seated on the left of the host two seats away, and the next in rank goes to the corresponding seat on the right.

Before the guests take their seats, chopsticks, spoons, wine cups, saucers and bone plates are arranged systematically in front of each seat. A set of these articles is supplied to each person, exclusively for his own use throughout the meal.

With the exception of four cold dishes which are placed on the table before the arrival of the guests, all the dishes are served one at a time, each new dish being placed in the centre of the table. As soon as the second dish is brought in, the first should be taken away. Present-day requirements call for the provision of a few extra spoons, which are placed in convenient positions on the table for the purpose of picking up food from the dishes in preference to one's own chopsticks. Two varieties of large spoons are used, the porcelain ones for soup and the silver ones for other food. This new arrangement meets with general approval, as, before its introduction, food used to be picked up from the main dish with the aid of chopsticks and transferred directly to the mouth.

When all the guests have arrived, the order to "warm" the wine is given. The wine commonly used in China is Shaohsing or rice wine, and it must first be warmed. When the wine is brought in, the host announces the name of the guest of honour, and fills his cup with wine, indicating his seat at the same time. This is repeated with the second and the third guests, and so on, until all the cups are filled except that of the host himself. To be polite the guest of honour should return the compliment by filling the host's cup.

When everyone is seated, the host takes his seat also. He then raises his wine cup saying *Ch'ing* (请) meaning "please". All the guests will then drink. After the drink the host again utters *Ch'ing* with a pair of chopsticks in his hand. The guest of honour then takes up his pair of chopsticks and begins to eat. His action will be followed by the other guests. Generally the host will wait until all his friends have had their share, as it is considered bad manners on his part to begin too soon. To enliven the proceedings the above ceremony may be repeated each time a new dish is served.

When the first main dish is brought in, usually shark fins, the host will request the guests to drain their cups saying *Kan Pei* (干杯), literally, "Dry cup", or simply *Ch'ing*. The guest of honour should now take the opportunity to thank the host for his hospitality.

If it happens to be a special occasion, such as a wedding banquet, and there are a large number of guests and many tables, the host will visit every table with a cup of wine in his hand and drink with the guests. The first main dish is served only in the presence of the host. This rite is called "presenting" (献). A big banquet has three presentings, but, as a rule, the ordinary wedding feast has only one. Guests take the opportunity to thank the host and drink with him immediately after the presenting ceremony. The return of compliment is called *Ch'ou* (酢). After this, due ceremony is considered to have been observed, and all are at liberty to enjoy the repast at leisure.

Should, however, as often happens, any of the guests show some hesitation in helping himself, the host will attempt to remedy the situation by saying *Sui Pien* (随便) which means "please do not stand on ceremony". Except on formal occasions a Chinese dinner is by no means a staid affair. The average Chinese is not devoid of a sense of humour, and can be counted upon not to miss any opportunity of contributing towards the "fun" of the evening. One of the ways of putting life into the party is by means of the finger game. No compulsion is employed: the loser is penalized by having to drink either a cup for each individual match, or a cup for the best of three matches. These matches may be played between any two guests, but on certain occasions the host plays with each guest in turn, and each guest should do likewise.

To the foreigner, it may seem strange that the loser, and not the winner, should have the pleasure of drinking, but, if careful reflection be given to the matter, it will be agreed that the Chinese are wise in that their viewpoint, in this as in many other instances, is diametrically opposed to the Western.

It is said that as soon as the wine is in, restraint is out, so once the flow of wine has been thus started, it is naturally to be expected to lead to a certain amount of conviviality and hilarity during the rest of the evening's proceedings. We often wonder whether foreign cocktails are intended to serve the same purpose.

Following the main dishes, which always end with a large bowl of duck soup and a fish, rice and pastries are served, and, at the same time, the supply of wine is cut off. As a matter of fact, the rice is seldom touched, and the reason is obvious. The fruit item, which is usually very welcome, is partaken of in an adjoining room. At this point hot towels are handed round—a Chinese custom which serves a double purpose in that it replaces the modern finger bowl, and acts, at the same time, as a signal to mark the end of a perfect evening.

The following rules governing tables manners are translated from Chapter 1 of the old Chinese classic—*Book of Etiquette*—written about 2,000 years ago.

DON'TS AT A CHINESE DINNER TABLE

1. 毋抟饭   Do not roll the rice into a ball.
2. 毋放饭   Do not bolt your food.
3. 毋流歠   Do not swill your soup.
4. 毋咤食   Do not eat audibly.
5. 毋啮骨   Do not crunch bones with your teeth.
6. 毋反鱼肉   Do not replace fish and meat which you have already tasted.
7. 毋投与狗骨   Do not throw bones to dogs.
8. 毋固获   Do not make a grab at what you want.
9. 毋扬饭   Do not spread out the rice to cool.
10. 毋嚃羹   Do not draw particles of food in the soup through your mouth; use chopsticks for this purpose.
11. 毋絮羹   Do not stir or add condiments to the soup in the common bowl.
12. 毋刺齿   Do not pick your teeth.

HOW TO HOLD CHOPSTICKS

Chopsticks are made use of to pick up morsels of food from a dish, and to aid in the transfer of rice from a bowl to the mouth. They are always in pairs, each stick being about 9 inches long, and they are so held in the hand that one is fixed in position while the other is movable. The correct way of holding them is as follows:

Hold the fixed one in your hand as in Figure 1 opposite, making it firm by pressure between the inside of the thumb and the end of the fourth finger. Then hold the other, the movable one, as you would hold a pen, but in a more upright position, as in Figure 2.

Use the middle finger to perform the movements necessary, the tip of the thumb acting as a pivot. In this way the two sticks can be used to act as a pair of pincers for taking up the food.

Figure 3 shows the correct way of holding the two sticks.

Table Manners 餐桌礼仪

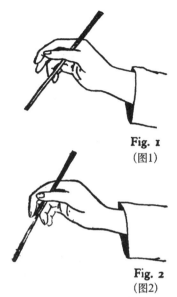

Fig. 1
(图1)

Fig. 2
(图2)

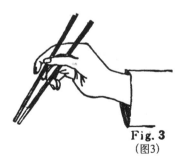

Fig. 3
(图3)

HOW TO HOLD RICE BOWLS

Hold the rice bowl with your left hand as shown in Figure 4. To get a firm grip the thumb should be placed on the upper rim and the first two fingers on the bottom of the bowl.

Figure 5 shows the position of the elbows while the right hand holds the chopsticks and the left hand holds the bowl.

Figure 6 indicates the correct position just at the point of eating. Certain definite rules are observed in regard to the manipulating of chopsticks and the rice bowl. Even the left-handed child is taught to hold his chopsticks with the right hand and the rice bowl in the left. On no account should the chopsticks be employed to shovel the rice from the bowl into the mouth; nor should the bowl be moved unduly. Only a gentle turn of the chopsticks by wrist action is needed to transfer the rice from the bowl to the mouth. The elbows should be kept as close as possible to the body, and should not be pointed outwards as they are likely to annoy your neighbours.

Fig. 4
(图4)

Fig. 5
(图5)

Fig. 6
(图6)

餐桌礼仪

餐桌礼仪的重要性再怎么强调都不为过。这些基本规则让来自不同社会和文化背景的人可以一边共享美餐一边交流互动。今天这些礼仪的作用与三千年前孔子《礼记》中所讲的并无二致:良好的礼仪如何向宾客展示尊重和快乐,如何将一个可能混乱无序的活动变为顺畅欢乐、令人舒服的体验。考虑到我们的社会已如此全球化,出国旅行和海外工作机会日益普遍,我们的餐桌上有来自各种文化传统和社会阶层的宾客,这些基本规则甚至更加重要了。良好的礼仪是一种国际通用的语言,让餐桌上的所有人都宾至如归。

徐芝韵

座位安排

《论语》有云:"为政以德,譬如北辰。居其所而众星共之。"古称帝王为"天子",其王位北坐而面南,犹如北辰。是以,中国素有以坐北朝南为尊之习俗。诸君排位列席时,以下情况亦当考虑进去:

"炕"可视为民屋之上座。"香案"则为放置供奉神明或祖先供品的高脚窄桌。房中若置有二者之一,则不论北方为何,以其所置处为北;若两者均无,则以入口处为南。

以圆桌飨宾客,客居北为尊,主则居南而面客。另,左席尊于右席。其渊源虽已难考,许是坐北朝南时旭日东升于左侧之故。主宾左侧为次席,右侧为三席,依次推之。故而与主人邻座者,或末宾,或亲友。

然则,南方商界,情形又有所不同。与主人正对而坐之人,被视作次主。主宾席列于主座左侧两席之隔处,次宾席于右侧之相应位置。

箸、勺、杯、碟和骨盘皆应于宾客入席前依序摆放上桌。餐具一人一套,仅为己用。

诸宾入座前,四盏冷盘先行置于桌上以待宾,其后则逐一进馔。新肴俱置于中央,待有后者奉上方移至旁处。因今时之人多有不以自用筷取食之习惯,故而尚需摆放数把公勺于趁手处,以便取食之用。公勺俱大,概分两类,瓷者舀汤,银者取馔。此布桌新仪乃应众需所生。此前,一应菜肴俱是以筷直取后入口。

诸宾俱至后,遂温酒。酒常取绍兴酒或米酒,温后方可呈上。酒即奉,主呼首宾姓名并示其席,斟酒至满。其后二宾、三宾乃至众宾皆按此序斟酒,仅余主杯为空。为感主人之谦恭,由首宾代为其斟酒。

众宾皆入座,主人乃入座,举杯言"请"(英文为please),诸宾遂饮。饮毕,主人举箸再言"请",首宾举箸以啖,众宾乃随。主人常待宾朋尽享后始啖,盖视促促自尝为失仪之举。为活跃气氛,每有新肴进馔,则复行此饮食之序。

及至主菜首呈,常为鱼翅,主人言"干杯"(英文直译为dry up)或简言"请",邀宾朋尽饮。首宾当此为机,尽表铭感主人款待之情。

若遇殊境,譬如婚筵,酒桌栉比而列,宾朋如云团坐,则主人当持杯遍访众桌,与诸宾共饮。首呈主菜须当主人之面,是为"献"礼(英文为presenting)。华筵当有三"献",然,寻常婚筵亦可作一"献"。献礼毕,诸宾遂举杯回敬主人,以酬谢意,是为"酌"。至此,应循之礼俱毕,众可自行乐食酣饮也。

然,遇宾客拘泥不能自乐之况,主当言"随便"以缓气氛,即"勿拘泥于场合"之意。除却正式场合,中式宴饮鲜有乏味之景。中国人通常非缺乏幽默之辈，亦非愿错失任何增添晚宴乐趣之机会。划拳即活跃聚会气氛一法。游戏规则无强制：或一局制胜，负者以酒自罚一杯；或三局二胜，自罚一杯。宾客两两结对即可行乐，然某些场合主人逐一与宾客结对，宾客亦须如此。

外邦之人许感于何以负者得饮酒之趣，而非赢者耶。然，细思之，可明中国人处世之妙处，亦如其他诸事，与西方截然不同尔。

酒水呈上，即开饮食之禁。兴敬酒之礼，引众享钟鸣鼎食、觥筹交错之趣。吾等亦常思，外邦筵席所伺鸡尾酒之目的，盖同矣。

鸭煲、鱼肴为主菜之末，其后呈米饭与糕点，是时，奉酒亦止。然则，米饭鲜有人尝，其故不言而明。备于侧厢内之瓜果奉上后，则颇受青睐。宴末之时奉热巾，此习俗有二用：一为净手，效同现代之洗指盅，二则标志宴毕。

下述餐仪则出自两千多年前所著古籍《礼记》之首篇。

中国餐仪之禁

1. 毋抟饭　勿搓饭成团而食。
2. 毋放饭　勿囫囵食之，放肆而无节制。
3. 毋流歠　勿大口灌汤。

4. 毋咤食　勿啧啧而食。

5. 毋啮骨　勿嚼骨。

6. 毋反鱼肉　勿将已食鱼肉放回食器。

7. 毋投与狗骨　勿投骨喂狗。

8. 毋固获　勿争食挑食。

9. 毋扬饭　勿贪速而扬饭凉之。

10. 毋嚃羹　勿贪饮而以口取食羹中之菜，当以箸取食之。

11. 毋絮羹　勿于公碗中添搅佐料。

12. 毋刺齿　勿恣意剔牙。

执　箸

箸为取食之用，碟中择菜，碗中渡米。箸身长约九英寸[1]，并成双以用，执箸时一动一不动。执箸之正姿如下：

不动者贴虎口，以拇指内侧及无名指指端之夹力固之（如第47页图1）。动者之执姿与握笔相类，稍竖之（如第47页图2）。

中指视需而动，拇指端之用则类轴。以此之法，二箸即可如钳物般取食。

执箸之姿见第47页图3。

1　1英寸合2.54厘米。——编者注。

持 碗

第 49 页图 4 示左手执碗之正确姿势。拇指压住上沿，食指中指抵住碗底。

第 49 页图 5 示右手执箸、左手持碗及双肘摆放之法。

第 49 页图 6 示进食时之正确姿势。据观，右箸左碗乃惯例，左利稚童亦授以此法。取食时，勿用箸铲米，恣意移动碗具亦不雅，凭腕力轻转箸即可取碗中米送入口中。双肘需贴靠身侧，切不可外张以致侵妨邻座。

TABLE SERVICE

A complete Chinese dinner service for ten persons consists of 148 pieces. This may be either of porcelain or silver, the latter being used only by wealthy families, while the porcelain is perhaps the more serviceable. Pewter articles were once used a great deal but they are now being rapidly displaced by the porcelain variety. Porcelain produced in Kiangsi Province is the best, because of the excellent quality of the clay available in the vicinity of *Poyang Lake* (鄱阳湖) where more than a dozen varieties can be found. The town of Ching Te Chen (景德镇), in Kiangsi, is responsible for nearly half of the porcelain in China. Its products are exported through the port of Kiukiang—hence they are called Kiukiang porcelains.

Ching Te Chen was one of the most important centres of the Chinese ceramic industry as far back as 200 A. D., and has since that date made the most beautiful china for the Imperial family. The famous "rice" pattern had its origin there. The porcelain made in Kwangtung Province (广东省) and exported through the city of Canton (广州) is known as Canton porcelain. This is, however, a grade inferior in quality to the Kiangsi product. There are two distinct styles of table crockery in use. The new style is thin, shallow and round in shape. The decorations consist generally of Chinese figures, flowers and birds, or landscapes. White on both surfaces is the common type, although the coloured variety is always obtainable. The old style of porcelain is thicker, deeper and usually hexagonal or octagonal in shape. The outer surface is usually dark blue or imperial yellow, and covered with antique Chinese designs, while the inner is generally of light blue colour.

Excellent copies of the old porcelain are now made in Kiukiang. These are extremely effective as table decorations and are much admired by foreigners.

Silver table sets are only seen in wealthy families. Each set consists of two wine pots and individual wine cups, soup spoons, pairs of chopsticks, a small dish for nuts or water-melon seeds, and another for soya bean sauce. In addition to these a tiny tray is provided for the wine cup and soup spoon, and a dainty rest for the chopsticks. A table decorated with beautifully coloured porcelain dishes and a well-made set of the above silver articles presents a display at once highly attractive and ornamental.

餐桌食具

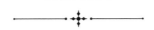

中餐十人之宴需备食具一百四十八件。瓷、银俱可，然后者仅见于富庶之家，前者更普众。彼时尝广用锡具，今已由瓷具迭代之。瓷者，以江西省产最佳，皆因该省鄱阳湖周边所产黏土质优，种类逾十二之故。中国近半瓷具出于江西景德镇，凭九江诸港之便利远销外界，故又称"九江瓷"。

"瓷都"景德镇乃中国瓷业之中心，其历史可追溯至公元二世纪，旧时即为官窑，专为皇家制供精品瓷器，闻名天下之"米通瓷"即源于此。广东省产又经广州销至海内外之瓷器，名曰"广州瓷"，然，其质稍逊于江西产。

陈于餐桌上之瓷器通具两种风格。胎薄、色浅、形圆者为新式，器身内外常以白为底色，并饰中国传统人物、花草、鸣禽或风景，然彩底者亦非难寻。胎厚、色深、形六角或八角者为老式，常以靛或明黄为外底色，饰以中国古典图样，内则常以月白为底色。

当下，老瓷仿品中之精品当属九江瓷。奉于餐桌之上赏心悦目，颇受海外宾客青睐。

银器仅见于富庶之家。每副餐具包括二酒壶，酒杯、汤勺、箸各一套，二碟（一盛坚果或西瓜子，一盛酱油）。此外，尚需为杯、勺各配一小托盘，箸配箸枕。餐桌之上，缤纷瓷器与精制银器相映成辉，顿可为餐桌增色不少。

TEA

Tea might correctly be termed the national beverage of China as it is so generally drunk by all classes, and the habit is one of very long standing.

It is the Chinese custom even at the present day to welcome a guest with a cup of tea and this is observed by the high and low alike. Good tea is of a clear colour, greenish or reddish, and has a slightly astringent flavour. The poor quality is very light in colour and bitter in taste. There are an indefinite number of varieties of China tea, with a wide range of prices. According to the method of curing, tea is divided into two main classes, viz: green and black tea.

Green tea leaves are dried and roasted as soon as they are picked, while the black variety is allowed to pass through a process of fermentation for a certain number of hours before softening and roasting. The best green tea leaves are picked before "grain rain" (谷雨) which falls about Easter time when they are still young. The leaves are tender and the aroma strong. The Province of Fukien produces the best green teas in China from a mountain called Wu-I. It is therefore called *Wu-I Ch'a* (武夷茶)[1]. *Lung Ching Ch'a* (龙井茶) is another well-known green tea. It comes from the lake city of Hang-chow in Chekiang, where chrysanthemum tea (杭菊) is also famous. Scented teas, such as Jasmine tea (茉莉茶) or *Hsiang P'ien* (香片) are much appreciated by foreigners.

Black tea is produced in many districts. The better known varieties are *Kee-mun* (祁门) and *Lu-an* (六安) from Anhui, *Wu-Loong* (乌龙) from Fukien, and *Po-erh* (普洱) from Yunnan. Yet another popular kind of black tea is *Pekoe* (白毫) which is a small leaf with a fuzzy surface and is marketed either scented or unscented.

The golden rule of making tea is to boil the water, not the tea. Put the leaves in a Chinese teacup, pour boiling water over them, and cover the cup with the lid. In a few minutes your cup of delicious tea is ready. Sugar and milk are never used in the Chinese style.

---

1 It is suspected that the author has confused the difference between green tea and oolong tea. The famous *Wu-I Ch'a* is oolong tea, which is a semi-fermented tea. Editor's note.

茶

茶，可谓中国之"国饮"，饮茶之人遍布各阶层，饮茶之史亦源远流长。

不论地位如何，中国人皆爱以茶迎宾，时至今日未有改。茶中佳者，茶汤色清，呈微绿或微红，香带微涩。次者则色近无，味苦。中国茶种之驳杂令人咋舌，其价格亦有天壤之别。依加工之法，概分二类：绿茶与红茶。

绿茶采摘后，即经杀青、炒青之序；红茶则需经数时发酵后，方行烘焙之序。谷雨前，即西方复活节左右采摘之绿茶为上乘，此时茶叶尚嫩，叶柔且馥郁。绿茶之首，产自福建武夷山，故名"武夷茶"[1]；另一颇负盛名之"龙井茶"则出自浙江杭州，此地出产的"杭菊"亦声名远播。而海外之人则更偏爱茉莉茶、香片之类的香茶。

红茶产地颇多，以安徽之"祁门茶""六安茶"，福建之"乌龙茶"及云南之"普洱茶"最负盛名。另有一名曰"白毫"之小叶红茶，爱之者亦甚，其叶上覆有绒毛，以有香和无香而分。

---

[1] 此处作者疑混淆了绿茶与青茶的区别。武夷茶之有名者为青茶，是一种半发酵茶。自明朝以后，绿茶生产已从武夷山地区逐渐消失。——编者注。

沏茶之道，关键在于经沸煮者为水，而非茶叶。置叶茶杯中，沸水倾覆之，上盖，数分钟后即可得一杯清香中国茶。且中式茶饮，从不行加糖、加奶之序。

WINE AND SONG

———— ✽ ————

No dinner is complete without wine, which brings joy and drives away depression, and makes the old feel young and the young still more youthful. Taken in moderation, it is undoubtedly an excellent stimulant. In the far north where the climate is cold and kaoliang abundant, Kaoliang wine, which is somewhat stronger than Russian vodka or English gin, is commonly drunk.

In Central China a milder drink known as Shaohsing wine—named after its producing centre in Chekiang—is very popular. It is the wine of China. It has another name *Hua Tiao* () meaning flower decoration, because the jars in which the wine is kept usually bear a floral decoration. When a girl is born, it is the common practice for the parents to make several jars of Shaohsing wine, the quantity depending upon the size of their purse, and keep them in a cool and dark room, until their daughter is married, so that on her wedding day, they will have at least some good old wine with which to entertain their guests.

In the South, where the climate is much warmer, a still milder drink known as *Liao Pan* (料半) meaning half strength, is commonly used. Stronger varieties, the double distilled (双蒸) and triple distilled (三蒸) are, however, obtainable. The flavouring of the wine is usually added afterwards, such as orange blossom (橙花) and green plum (青梅).

Though Confucius was very particular in regard to food, his list of undesirable foodstuffs being a long one, he was not so with wine, as apparently all wines were acceptable to him. There are no records as to his capacity, but he is believed to have been a good drinker like the rest of those ancient scholars.

My own capacity does not exceed one wine cup of Shaohsing, so I am really not qualified to say much on the art of drinking. Perhaps it is well, for it is a subject in which I suspect our Western friends can give Eastern folk quite a few lessons.

In days gone by, good wine was named after an official in Chingchow. By a strange coincidence, the best Shaohsing wine obtainable at the wine merchant Wong Yu Ho in Shanghai is called LL. D. wine, after a famous Chinese lawyer who was one of his best customers. To secure some of this brand for my own use I have to go to the shop armed with a note from my legal friend as I know full well that I would not get it otherwise; the shop will not supply it to anyone who they think will not appreciate it.

The lawyer is now in Geneva, but I do not doubt that this shop continues to supply him with his favourite wine.

According to our ideas tea should be drunk in quiet surroundings, while wine should be accompanied by song. This may be the reason why restaurants are always noisy.

The renowned poets of old were, as a rule, good drinkers. Among them I may mention *Li Po* (李白), *T'ao Tsin* (陶潜), and *Pa Chu I* (白居易). Wine gave them inspiration, and, when they drank enough, they wrote beautiful verses, a great number of which are still recited by school boys and girls of today.

Below is a song composed by a well-known tippler of the past who used to cheat his wife to get liquor. One day he told his wife that he had made up his mind to give up drinking, but as he had had the habit for so long, it was only fair to allow him one final session before quitting. His wife was overjoyed to hear this, and proceeded to search the whole town for the necessary wine. When he got the wine, he immediately drank the entire lot in one gulp, and then sang the following verse referring to Chinese wine measures:

Liu Ling, Liu Ling,	天生刘伶
(that's my name)	
From drinking comes my fame.	以酒为名
A "hu" each bout I take,	一饮一斛

Five "tou" I need to wake. 五斗解酲

My wife she tries to plead, 妇人之言

Her words are naught to heed. 慎不可听

酒与歌

━━━━◆◆◆━━━━

无酒不成宴。酒可助兴舒压，使长者重焕青春，年轻者愈发朝气。适度而饮，乃绝佳助兴之物。北方偏寒之地盛产高粱，高粱酒之烈更甚于俄罗斯伏特加与英伦杜松子，饮者甚众。

华东之绍兴酒稍温和，十分流行，可谓国酒。因其产地乃浙江之酒乡绍兴，故得名绍兴酒；其又名"花雕"，即以花雕饰之意，皆因盛酒之坛壁上常饰有花形之故。民间有俗，家中得女，双亲即制绍兴酒数坛，长置于阴凉避光之所。而酒之数量，全仗富贫之况。待女儿结秦晋之好，便可在大喜之日奉上数坛陈年佳酿以乐宾客。

华南之地，气候更暖，惯饮之酒"料半"亦更温和。料半，即中度之意。然，"双蒸""三蒸"更烈之，亦可得。又有"橙花""青梅"等香酒，则为后添香料加工之酒。

孔子对入口之物素讲究食精脍细，不食之物甚多，唯对酒或酒类无框定。其酒量虽无文献可考，盖同多数古时文人一样擅饮。

吾之酒量，仅抵一杯绍兴酒，是以，难言深味酒之趣。此或非憾事，论起饮酒之道，吾怀疑西方友人比起东方民众亦不遑多让。

清朝时期，曾有以官员姓名为佳酿冠名之俗。无独有偶，吾从上海酒商黄宇浩（音译）处得来的上好绍兴酒即取名LL.D.酒，名从中国一位著名律师，而此人亦是黄的最佳顾客。为得此佳酿，吾须携律师好友之便笺访问酒家，而对不懂如何品鉴此酒的顾客，店家一概谢绝出售。

该律师现已在日内瓦，但吾丝毫不怀疑店家会继续为其供应他最爱之佳酿。

在中国人看来，饮茶当处清幽之境，饮酒则应以歌助兴。此大抵为餐馆总是人声鼎沸之根源。

名留青史之文人骚客，如李白、陶潜、白居易等，无不是擅饮之辈。杯中之物乃灵感之引，饮至酣处，诗兴奔涌，翰墨淋漓。时至今日，学生们依旧在朗朗背诵他们的众多佳作。

以下诗歌乃"竹林七贤"之一亦是著名酒鬼的刘伶，为哄骗妻子以获贪杯之乐而作。某日，他告知妻，言戒酒心已定，奈何酣嗜时日已久，当在戒酒之前痛饮最后一回。妻子闻言，欢喜异常，为其遍访村镇寻酒。然，刘伶得酒后，便一口饮尽，遂吟出这首与中国酒量之法有关的诗歌：

> 天生刘伶，
>
> 以酒为名。
>
> 一饮一斛，
>
> 五斗解酲。
>
> 妇人之言，
>
> 慎不可听。

KITCHEN UTENSILS

1. Rolling Board	Kan Mien Pan
2. Large Rolling Pin	Ta Kan Mien Kun
3. Small Rolling Pin	Hsiao Kan Mien Kun
4. Chopper	Ts'ai Tao
5. Chopping Block	Ts'ai Tun Tzu
6. Meat Slice	Ch'an Tzu
7. Bamboo Chopsticks	Chu K'uai Tzu
8. Saucepan	Ta Kuo
9. Steam Cooker	Chêng Lung
10. Small Strainer	Hsiao Lou Shao
11. Ladle	Shou Shao
12. Large Strainer	Ta Lou Shao
13. Chinese Frying Pan	Ch'ao Shao
14. Deep Earthenware Saucepan	Sha Kuo

Kitchen Utensils 炊具

注：图中"趕"今写为"擀"，"麵"今写为"面"，"墊"今写为"墩"，"鏟"今写为"铲"，"焗"今写为"锅"，"籠"今写为"笼"，"沙"今写为"砂"。——编者注。

炊 具

1. 擀面板
2. 大擀面棍
3. 小擀面棍
4. 菜刀
5. 菜墩子
6. 铲子
7. 竹筷子
8. 大锅
9. 蒸笼
10. 小漏勺
11. 手勺
12. 大漏勺
13. 炒勺
14. 砂锅

INGREDIENTS AND CONDIMENTS

SOYA SAUCE

Soya sauce plays an important part in Chinese cooking as it imparts a flavour and taste totally different from salt.

In order to get the best flavour and taste it is always advisable to use the best quality.

SHAOHSING WINE

When using wine in cooking it is best to use *Shaohsing* Wine. If not obtainable, sherry can take its place. It is never advisable to use wine which has turned sour. To preserve its flavour, keep in a cool place.

WINTER MUSHROOMS

How to soak dried mushrooms [*Tung Ku* (冬菇)]:

Wash the necessary amount of mushrooms thoroughly two or three times, and soak them in boiling water for 15 minutes. Then pick off the stems, and they are ready for use.

HOW TO BOIL MUSHROOMS

Take 1/2 lb. of soaked mushrooms, and cook them in a deep saucepan with 3 cups of cold water, 3 teaspoonfuls of soya sauce, 1 teaspoonful of salt, 2 teaspoonfuls of sugar and 1/2 lb. of pork, for 10 minutes over a fast fire, and allow to simmer for 50 minutes. The mushrooms and sauce are then ready for use in such recipes as they are required for.

Three kinds of mushrooms:

1. Straw mushrooms *Ts'ao Ku* (草菇) grow from dried straws of glutinous rice. They are black in colour, and are usually cut into two halves and dried.

2. Winter mushrooms grow from wood. The thin kind is called *Tung Ku* (冬菇) and the thick *Hua Ku* (花菇). They are dark brown in colour when dried. Fresh mushrooms of this kind are very seldom seen in the market.

3. Button mushrooms can be obtained fresh or dried. They are called *K'ou Mo* (口蘑), and are excellent for soup making.

Ingredients and Condiments　　配料与调料

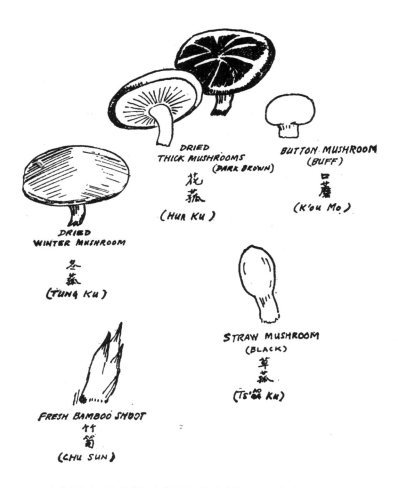

注：图中"菰"今写为"菇"，"笋"今写为"笋"。——编者注。

RED HAWS [*SHAN CHA PING* (山楂饼)]

Used for Sweet and Sour Pork, etc., according to the recipes, are crab apples, with the cores taken out, crushed into thin round cakes after addition of sugar, and dried. These can be procured and, being in the dry form, easily stored.

BAMBOO SHOOTS

Tinned bamboo shoots can be used for cooking as they are, but fresh ones must be first boiled in water for about 10 to 15 minutes before use. The water should be discarded.

GINGER

For the following recipes, when ginger is used in tiny slices, it is measured by the tip of a finger; when used in big slices, by the tip of the thumb.

CHINESE ONION

The word stalk used in the recipes refers to about 2 inches only, and not the full length. When minced or finely chopped, it means a length of about 1/4 inch.

Ingredients and Condiments 配料与调料

For those resident in America or in England who are desirous of securing ingredients for Chinese cooking, the numerous Chinese restaurants now existing in those countries should be able to supply them on request. If not obtainable, the following substitutes may be used:

Sweet and sour pickles may be substituted by foreign pickles.

Sesamum oil may be substituted by "Wesson oil".

Red haws may be substituted by crab apple jelly.

Star aniseed and red pepper may be substituted by bay leaves.

The measurements, such as cups, teaspoons, etc., mentioned in the following recipes, are exactly the same as those used in foreign countries.

配料与调料

酱油

酱油乃中式料理烹饪重要之物,其风味独特,与盐截然不同。

为得最佳风味与口感,建议取质最优者用。

绍兴酒

料酒中以绍兴酒为最佳,无可得时,可以雪莉酒代之。如遇料酒已变质发酸者,不建议使用。为保料酒之醇香,需置于阴凉处。

冬菇

浸渍干菇之法:

依需取适量干菇洗濯二到三次,后浸于沸水中十五分钟。去茎干,即可用矣。

蘑菇汤

煮蘑菇汤之法：

取浸渍好的蘑菇二分之一磅，置于炖锅中，添冷水三杯、酱油三茶匙、盐一茶匙、糖二茶匙、猪肉二分之一磅，高火烹煮十分钟后，文火炖煮五十分钟。其后，蘑菇与汤汁即可因需用于菜肴烹饪。

三种常用蘑菇：

1. 草菇：生于干燥糯米秆堆上，色深，常一切为二，曝后用。

2. 香菇：生于树干之上，瘦为冬菇，厚为花菇，曝后呈深赭。卖场罕见新鲜者。

3. 洋菇：鲜、干二者均用，又名口蘑，是为炖汤之绝佳材料。

山楂饼

烹饪糖醋排骨等用。制法：山楂去核，加糖，压制成圆薄饼后曝干。此法可使山楂饼便于买卖，亦利于保存。

竹笋

罐装竹笋开罐即可烹饪，鲜笋须焯水十至十五分钟后方可用，焯后水弃之。

姜

本作食谱中之姜丝，细者以指尖量之；粗者以拇指尖量之。

葱

本作食谱中之葱段，非指整棵葱，而是切成约二英寸长。如是剁碎或切成小段，则约四分之一英寸长。

对居于美英而冀望觅得中式食材之诸位，而今遍布各城市之中餐馆或能提供此便利。如若不然，可参考以下替换物：

糖醋泡菜可以外国酱菜代之；

芝麻油可以威臣沙律油代之；

山楂饼可以山楂酱代之；

八角茴香、红辣椒可以月桂叶代之。

本作食谱提及之计量法，如杯量、匙量等，概与国外计量法无异。

Selected Recipes

中饡珎𦙫

Selected Recipes 中馈珍藏

BASICS
[Chi Ch'u Lei (基础类)]

Pages

84 How to Make a Cup of Tea

86 How to Cook Rice

88—89 Fried Rice

基础类

85 香片茶的泡法

87 米饭的煮制法

90—91 什锦炒饭

HOW TO MAKE A CUP OF TEA
[Hsiang P'ien (香片)]

Proportion:

    1   coffee-spoonful of tea leaves
    1   measuring cup of boiling water

Method:

Put the leaves into a Chinese teacup. Then pour on water that has just boiled, and cover with the lid. Allow it to stand for 3 minutes, without stirring. It is then ready for serving.

香片茶的泡法

材料：

　　一小匙茶叶

　　一茶杯开水

做法：

　　首先将茶叶放在中国茶杯中，然后把刚开的水冲上，再用盖子盖上，让它停三分钟，不需动它，之后就可饮之。

HOW TO COOK RICE
[Shao Fan (烧饭)]

Proportion:

    1 Chinese bowl of rice
    2 Chinese bowls of water

Method:

Wash the rice thoroughly 4 or 5 times in cold water. Put the rice in a deep saucepan, add the water, bring to boiling point, and let it boil for 7 minutes. Then simmer over a small fire for 30 minutes.

The above will make 3 bowls of rice.

米饭的煮制法

材料：

一饭碗大米

二饭碗水

做法：

将大米用冷水洗上四至五次，放进深点的平底锅，加上水，让它沸至七分钟，然后用小火候煮三十分钟。这样做法做的饭是很芬香可口的。这分量可做出三碗白饭。

FRIED RICE
[Shih Chin Ch'ao Fan (什锦炒饭)]

Ingredients:

- 8   bowls of cold cooked rice
- 1   onion (finely chopped)
- 1/4   lb. cooked ham (finely chopped)
- 8   mushrooms (previously soaked and finely chopped)
- 1/4   lb. chicken meat (cooked and cut into small pieces)
- 3   whole eggs (beaten up and fried until they are quite firm, then broken up into small pieces)
- 1/4   lb. tinned green peas
- 1/4   lb. shrimps (cooked)
- 1/4   lb. tinned bamboo shoots (cut into squares)
- 1/4   lb. Chinese sausage (cooked and cut into squares)
- 5   teaspoonfuls of soya sauce
- 1/2   cup of lard
- 1   teaspoonful of salt

Method:

Put 1/4 cup of lard into a hot frying pan with 1 teaspoonful of salt. Add the rice and fry for 5 minutes. Then turn out into a bowl.

Add another 1/4 cup of lard. Fry the onion until a light brown, then add the remaining ingredients and fry for 3 minutes.

Now add the rice and 5 teaspoonfuls of soya sauce, and mix thoroughly.

Serve hot in a large bowl. The above quantity is sufficient for 8 persons.

什锦炒饭

材料：

八小碗冷白饭

一只洋葱，切碎

四两[1] 熟火腿，切碎

八只冬菇，泡发，切碎

四两煮熟的鸡肉，切成丁

三只鸡蛋，炒熟，弄碎

四两罐装青豆

四两虾仁，煮熟

四两罐装竹笋，切丁

四两熟香肠，切丁

五小匙酱油

半杯猪油

一小匙细盐

---

1 本书用的是中华民国年间的"两"，一两约合 31.25 克。——编者注

做法：

将二两猪油倒置热炒锅中，加入一小匙细盐，再把白饭倒入，约炒五分钟，然后把它盛在大碗里备用。

现在将余下二两猪油倒在炒锅里烧热，将洋葱炒成金黄，再加入其他全部材料炒三分钟，再将已炒过的白饭倒进，加入五小匙酱油混合炒之，顺和后就可以供食用了。这个量可供八位食之。

FOWL DISHES
[Ch'in Lei (禽类)]

Pages

94	Gold Coin Chicken
96	Walnut Chicken
98	Chili Oil Chicken and Spinach
100	Stewed Chestnut Chicken
102	Fried Chicken with Pepper and Brown Sauce
104	Velvet Chicken (Imitation)
106	Roast Stuffed Chicken
108	Roast Chicken (Boneless)
110	Velvet Chicken with Corn
112—113	Stuffed Mushrooms
116	Pineapple and Ginger Duck
118	Roast Crisp Duck
120	Red Sauce Duck
122—123	Fried Duck Liver
126	Fried Wild Duck
128—129	Minced Pigeon

禽 类

95	金钱鸡
97	核桃鸡丁
99	辣油鸡丁
101	栗子焖鸡
103	酱泡鸡丁
105	芙蓉假鸡片
107	烤酿鸡
109	锅烧鸡
111	鸡蓉玉米
114—115	酿冬菇
117	菠萝姜鸭子
119	脆皮鸭子
121	红烧扒鸭
124—125	炒鸭肝
127	炒水鸭片
130—131	鸽子松

GOLD COIN CHICKEN
[Chin Ch'ien Chi (金钱鸡)]

Ingredients:

1/4	lb. chicken fillet (sliced)	⎫
1/2	teaspoonful of cornflour	⎬ mix together
1/2	teaspoonful of salt	⎬
3	teaspoonfuls of wine	⎭

- 1/4 lb. roast pork (sliced)
- 1/4 lb. cooked ham (sliced)
- 1/4 lb. cooked mushrooms (*Tung Ku*) (whole)
- 1/3 cup of lard

Method:

Transfix on a skewer, about the length of a pencil, pieces of chicken, ham, mushroom and roast pork, one piece of each at a time, in the order mentioned. The above material is sufficient for 4 such skewers.

Rub them well over with the lard, then toast them over an open fire for 15 minutes as you would toast bread, or roast them in a quick oven for 15 minutes. Remove skewers before serving.

金钱鸡

材料：

四两鸡肉，切片

半小匙玉米粉

半小匙细盐

三小匙米酒

以上四种材料混合在一起。

四两叉烧肉，切片

四两熟火腿，切片

四两熟冬菇，用整个的（大约要一两干的冬菇）

三分之一杯猪油

做法：

用像铅笔那么长的烧叉将以上的鸡片、火腿、冬菇、叉烧肉等材料一一地刺穿起来，分成四支烧叉将它装上。

让它干一下，淋上猪油，然后将它烘之，约十五分钟之久，好像你在炭火上烘面包一样操作；或者用强火候的烤箱来烤十五分钟。熟后再将烧叉除去，食之。

WALNUT CHICKEN
[Hê T'ao Chi Ting (核桃鸡丁)]

Ingredients:

- 1 lb. chicken fillet (cut into squares)
- 1 cup of walnuts
- 1 slice of ham (cut into squares)
- 1/2 cup of lard
- a pinch of salt
- 2 teaspoonfuls of wine ⎫
- 1/2 teaspoonful of salt ⎬ mix in chicken
- 1/2 teaspoonful of cornflour ⎭

Method:

Blanch the walnuts in boiling water for 15 minutes, adding a pinch of salt. Take them out, remove the skin and fry them in deep fat until light brown. Allow to cool.

Now fry the chicken in 1/2 cup of lard for 1 minute. Remove and drain.

Using the lard left in the pan fry the chicken again with the walnuts and ham for 1 minute, then serve.

核桃鸡丁

材料:

一磅鸡肉,切丁　　二小匙米酒

一杯核桃肉　　　　半小匙细盐

一片火腿,切丁　　半小匙玉米粉

半杯猪油　　　　　以上三种材料混合在鸡肉里。

少许细盐

做法:

将核桃肉用开水泡上十五分钟,加入少许细盐,然后去皮用油炸之,成金黄色盛出,让它冷之。

现在将半杯猪油烧热,把鸡肉略炒一分钟,倒出控油。

用锅里剩下的猪油把炒过的鸡肉、核桃肉和火腿再炒一分钟,即食之。

CHILI OIL CHICKEN AND SPINACH
[La Yu Chi Ting (辣油鸡丁)]

Ingredients:

3/4	lb. chicken fillet (cut into squares)	⎫
1/2	teaspoonful of salt	⎬ mix together
1/2	teaspoonful of cornflour	
1½	tablespoonfuls of water	⎭
4	hot chilis (cut into halves)	
1	lb. spinach	
1	teaspoonful of wine	
1	teaspoonful of soya sauce	
1/2	teaspoonful of salt	
3/4	cup of lard	

Method:

Heat 1/2 cup of lard in a frying pan, fry the hot chilis until brown, and remove them.

Using the same fat fry the chicken. Add the wine and soya sauce.

Fry the spinach with 1/4 cup of lard and salt for 2 minutes.

Serve in a dish, the chicken on one half of the dish, and the spinach on the other, after garnishing with a little of the fried chilis chopped up.

辣油鸡丁

材料：

四分之三磅鸡肉，切丁

半小匙细盐

半小匙玉米粉

一又二分之一大匙水

将以上的材料混合在一起。

四只小辣椒，切成对开

一磅菠菜

一小匙米酒

一小匙酱油

半小匙细盐

四分之三杯猪油

做法：

将半杯猪油倒进炒锅中烧热，把辣椒炒成发黄，把它盛出。用那些油炒那鸡肉丁，加酒和酱油。

用四分之一杯的猪油置在炒锅中烧热，将菠菜加盐略炒之约二分钟。

将鸡丁放置于盘子的一边，再把菠菜放在另一边，将那炒过的辣椒切碎散在上面，看起来很美观，这真是色香味俱全。

STEWED CHESTNUT CHICKEN
[Li Tzu Mên Chi (栗子焖鸡)]

Ingredients:

1	spring chicken (about 2 lbs., chopped with bone)
1	cup of chestnuts (cut into halves and boiled 15 mins.)
1	slice of ginger
1	small stalk of Chinese onion
1	teaspoonful of salt
6	teaspoonfuls of soya sauce
1	teaspoonful of sugar
1	teaspoonful of wine
1/4	cup of lard
2	cups of water

Method:

Heat the lard in a frying pan and fry the ginger and onion, then add the chicken, soya sauce, salt, sugar, wine and water. Then simmer for 3/4 of an hour.

Now add the boiled chestnuts and simmer for another 15 minutes. Serve with gravy.

栗子焖鸡

材料：

一只约二磅重的春鸡，连骨切成块

一杯栗子肉，切成两半煮十五分钟

一片生姜

一小根青葱

一小匙细盐

六小匙酱油

一小匙白糖

一小匙米酒

四分之一杯猪油

二杯开水

做法：

将油置在炒锅中烧热，把生姜、青葱炒数下，再把鸡块倒进，然后加入酱油、细盐、白糖、酒、水，用文火焖煮四十五分钟。

现在加入那已煮过的栗子肉，再用慢火焖十五分钟。食时连汁同上之。

FRIED CHICKEN WITH PEPPER AND BROWN SAUCE
[Chiang P'ao Chi Ting (酱泡鸡丁)]

Ingredients:

- 3/4 lb. chicken fillet (cut into squares)
- 2 mushrooms (soaked and cut into squares)
- 2 green peppers (cut into squares)
- 2 red chilis (cut into squares)
- 1 bamboo shoot (cut into squares)
- 3 teaspoonfuls of chiang (甜面酱, a thick brown sauce)
- 3/4 cup of lard
- 1/2 teaspoonful of salt
- 2 teaspoonfuls of wine
- 1/2 teaspoonful of cornflour

Method:

Mix the chicken with salt, cornflour and wine well.

Heat 1/2 cup of lard in hot frying pan. Fry the chicken for 2 minutes. Take out and drain.

Add 1/4 cup of lard and fry the chiang a little. Fry the mushrooms, peppers, chilis and bamboo shoot together for 2 minutes. Then replace the chicken and fry another minute. Serve very hot.

酱泡鸡丁

材料：

四分之三磅鸡肉，切丁

二只冬菇，用开水泡发，切丁

二个青辣椒，切丁

二个红辣椒，切丁

一只竹笋，切丁

三小匙甜面酱（一种浓稠的褐色调味汁）

四分之三杯猪油

半小匙细盐

二小匙米酒

半小匙玉米粉

做法：

首先在鸡丁中加入细盐、玉米粉、米酒等，混合之。

倒入半杯猪油在炒锅中，烧热之后将鸡丁爆炒二分钟，把鸡丁盛出。

用四分之一杯猪油炒那甜面酱，再加入冬菇、青辣椒、红辣椒、竹笋等约炒二分钟，然后将那鸡丁倒入再爆炒一分钟，盛起即供食之。

VELVET CHICKEN (IMITATION)
[Fu Jung Chia Chi P'ien (芙蓉假鸡片)]

Main Ingredients:

- 10 egg whites
- 1 lb. spinach
- a little shredded bamboo shoot
- 1 slice of cooked ham (finely chopped)
- 1 teaspoonful of salt
- 1¼ cups of lard

Sauce Ingredients:

- 1/2 teaspoonful of salt
- 1/2 cup of stock
- 2 teaspoonfuls of cornflour

Method:

Stir the egg whites up a little with the salt. Heat the lard moderately in a pan, pour in the eggs and let it set. Remove and drain.

Leaving a cup of lard in the pan, fry the vegetables, and when they are cooked, drain away the liquid and turn out onto a dish.

Put the stock in a hot pan. Add the salt and the cornflour mixed with a little water.

To this sauce add the cooked egg and stir for a second. Pour this on top of the vegetables. Garnish with the ham and serve.

芙蓉假鸡片

主菜的材料：

十只鸡蛋的白　　　　　　　一片熟火腿，切碎（约二两）
一磅菠菜　　　　　　　　　一小匙细盐
少许竹笋薄片　　　　　　　一又四分之一杯猪油

调味汁的材料：

半小匙细盐　　　　　　　　二小匙玉米粉
半杯原汁肉汤

做法：

用一些细盐将蛋清打之。把猪油置在炒锅中烧得温热，然后把已调顺的蛋倒进，用文火烧一会，至发白时，将已熟的蛋盛起。

那余下的油留在炒锅中，大约还有一杯，再炒那蔬菜。烧好之后的菜会有汤，把它倒掉，将菜装在盘上。

把原汁肉汤烧热，加入细盐，用少许水将玉米粉调和加入。

小沸时将已煮好之蛋倒进，略翻数下，即离火把它倒在蔬菜上面，然后再把那切碎的火腿散在上头，这种装饰是很悦目的。

ROAST STUFFED CHICKEN
[K'ao Jang Chi (烤酿鸡)]

Ingredients:

1	spring chicken (about 2 lbs., boned)
1	lb. pork (minced)
1	slice of ginger (finely chopped)
1	stalk of Chinese onion (finely chopped)
1	heaped teaspoonful of cornflour
1/2	cup of water
1	teaspoonful of wine
1	teaspoonful of salt
1/2	teaspoonful of sugar
3	teaspoonfuls of soya sauce
1/4	cup of lard

Method:

Mix well together the pork, ginger, onion, cornflour, wine, salt, sugar, 2 teaspoonfuls of soya sauce and 1/2 cup of water. Put this stuffing into the boned chicken.

Roast in a hot oven with the lard and 1 teaspoonful of soya sauce for 3/4 hour.

Cut into big slices before serving.

烤酿鸡

材料：

一只春鸡（约二磅），去骨

一磅猪肉，搅细

一片生姜，切碎

一根青葱，切碎

一满匙玉米粉

半杯水

一小匙米酒

一小匙细盐

半小匙白糖

三小匙酱油

四分之一杯猪油

做法：

先将搅细的猪肉和生姜、葱、玉米粉、米酒、细盐、白糖、二小匙酱油以及半杯冷开水等材料调得很顺和，再将这些作料装在去骨的鸡的里面。

鸡的外皮擦猪油和一小匙酱油，用很热的烤箱（约两百摄氏度）来烤四十五分钟。

食的时候把它切成大块供餐。

ROAST CHICKEN (BONELESS)
[Kuo Shao Chi (锅烧鸡)]

Ingredients:

- 1 spring chicken (about 2 lbs.)
- 3 teaspoonfuls of wine
- 5 teaspoonfuls of soya sauce
- 3 teaspoonfuls of cornflour

Method:

Boil the chicken for 1½ hours, then bone it, and let it stand until cold.

Make a paste with the cornflour, wine and soya sauce. Rub this over the chicken, and fry it in deep fat until brown and crisp. Serve immediately.

锅烧鸡

材料：

　　一只春鸡（约二磅）

　　三小匙米酒

　　五小匙酱油

　　三小匙玉米粉

做法：

　　将洗净的鸡用开水煮一个半小时，然后盛出去骨，冷之。

　　用玉米粉、酒、酱油调成糊，抹在鸡的外皮上。需很多的油烧得很热，将鸡放进炸之，成金黄色且酥脆时盛起，即刻供食。这道菜大家都喜爱吃。

VELVET CHICKEN WITH CORN
[Chi Jung Yü Mi(鸡蓉玉米)]

Ingredients:

- 2   chicken breasts
- 1   small piece of fat pork
- 1   tin of sweet corn
- 1   slice of cooked ham (finely chopped)
- 1   teaspoonful of salt
- 1   teaspoonful of wine
- 3¾   cups of chicken stock
- 1   teaspoonful of lard
- 3   teaspoonfuls of cornflour

Method:

Mince the chicken and pork well together. Add 1/2 teaspoonful of salt, wine, cornflour and 3/4 cup of stock.

Put the rest of the stock and salt in a pan, and boil. When boiling, add the sweet corn and the chicken and pork mixture, then bring to the boil again, stirring constantly for about 3 minutes. Lastly, add the lard, and stir thoroughly.

Garnish with the ham, and serve in a deep bowl.

鸡蓉玉米

材料：

二片鸡胸肉

一小块肥猪肉

一罐甜玉米粒

一片熟火腿，切碎

一小匙细盐

一小匙米酒

三又四分之三杯原汁鸡汤

一小匙猪油

三小匙玉米粉

做法：

将鸡肉和猪肉切碎混合之，再加入半小匙细盐、米酒、玉米粉和四分之三杯原汁鸡汤。

把其余的原汁鸡汤倒入锅中，加些细盐煮之。将汤煮开之后，把那玉米粒和已调好之鸡与猪肉加入再煮，不断地调三分钟。最后，加入猪油，彻底搅拌。

食的时候把已切碎的火腿散在汤上面，装汤的碗需要大点。

STUFFED MUSHROOMS
[Jang Tung Ku (釀冬菇)]

Ingredients:

16	mushrooms (soaked)
2	chicken breasts (minced)
1/4	lb. pork fat (minced)
1	egg white
1	slice of cooked ham (finely chopped)
1½	teaspoonfuls of salt
3/4	cup of meat stock
3	teaspoonfuls of soya sauce
1	teaspoonful of sugar
2	teaspoonfuls of wine
2	teaspoonfuls of cornflour
1	teaspoonful of lard

Method:

Add to the chicken the pork fat, 2 teaspoonfuls of wine, 1/2 teaspoonful of salt, 1 teaspoonful of cornflour, the egg white, ham and 1/4 cup of meat stock, and mix thoroughly.

Cook the mushrooms in sufficient water. Add the soya sauce, sugar and 1/2 teaspoonful of salt, and simmer for 1 hour. Take out the mushrooms and allow to cool.

Now stuff each mushroom with the chicken filling, place in a double boiler and steam for 10 minutes. Serve with the sauce as made below.

Put the rest of the meat stock in a frying pan and add 1/2 teaspoonful of salt. Mix the cornflour with a little water in a cup. Add this slowly, together with the lard.

酿冬菇

材料：

十六只冬菇，泡发

二片鸡胸肉，切碎

四两肥猪肉，切碎

一只鸡蛋的白

一片熟火腿，切碎

一又二分之一小匙细盐

四分之三杯原汁肉汤

三小匙酱油

一小匙白糖

二小匙米酒

二小匙玉米粉

一小匙猪油

做法：

鸡肉里面加猪肉、二小匙酒、半小匙细盐、一小匙玉米粉、蛋白、火腿，和四分之一杯原汁肉汤全部混合在一起。

用足够的水将冬菇煮之，加入酱油、白糖、半小匙盐，用文火慢慢地煮上一个小时，然后把冬菇盛起让它冷之。

现在可将鸡肉装在冬菇内，全部装好之后把它蒸十分钟，食的时候用碗装，连如下制作之汤汁同餐：

其余的肉汁汤倒入锅中，加进半小匙盐，用一点水将玉米粉调成糊顺顺地加入，再加点猪油用慢火煮一会，随供餐。

PINEAPPLE AND GINGER DUCK
[Po Lo Chiang Ya Tzu (菠萝姜鸭子)]

Main Ingredients:

- 1 spring duck (about 3 lbs.)
- 1 small tin of pineapple
- 6 pieces of ginger (tinned)
- 1 teaspoonful of salt

Sauce Ingredients:

- 1½ teaspoonfuls of cornflour
- 1 cup of pineapple juice (from the tin)
- 1/2 cup of ginger juice (from the tin)

Method:

Steam the whole duck with salt for 2½ hours, then remove and allow to cool.

Now cut it up into large slices, and arrange these in the centre of a big dish. Also cut the pineapple and ginger into thick slices and arrange them alternately round the duck.

Heat the pineapple and ginger juice in a frying pan for a little while, then add the cornflour mixed with a little cold water to thicken it. Pour this on top of the duck before serving.

菠萝姜鸭子

主菜的材料：

一只春鸭（约三磅）　　　　　六片罐装生姜

一小碗罐装菠萝　　　　　　　一小匙细盐

调味汁的材料：

一又二分之一小匙玉米粉

一杯菠萝汁（罐中的）

半杯姜汁（罐中的）

做法：

将鸭子和盐蒸二个半小时，然后取出让它冷之。

现在将鸭子切成长块，装在大盘中央；同时把菠萝和生姜切成长条片交替排列在盘子的周围。

将菠萝汁和姜汁置在锅中烧热，然后把玉米粉用一点水调和，慢慢地加入汤汁中调成薄糊，吃时把它淋在鸭子上面餐之。

ROAST CRISP DUCK
[Ts'ui P'i Ya Tzu (脆皮鸭子)]

Ingredients:

1	duck (about 3 lbs.)
1	stalk of Chinese onion
1	small clove of garlic
1	big slice of ginger
3	small pieces of aniseed
5	teaspoonfuls of soya sauce
10	teaspoonfuls of salt

Method:

Cover the duck with cold water in a deep saucepan. Add the ingredients, and bring to the boil. Then simmer for 1½ hours.

Remove the duck. When cold fry it in deep fat for about 6 minutes, keeping it well basted, till a golden brown.

Serve while very hot and crisp. Either serve whole, or cut it up before serving.

脆皮鸭子

材料：

一只鸭子（约三磅）

一根青葱

一粒蒜头

一大片生姜

三小粒八角

五小匙酱油

十小匙细盐

做法：

将鸭子置在锅中，所需的水和鸭子平。然后加入那其他全部材料，煮沸。再用慢火候煮上一个半小时后，将它盛起，让它冷之。

再用很多油把它炸成金黄色，大约需六分钟。

这道菜是出锅马上供餐，你可以整个上桌或切成块。

RED SAUCE DUCK
[Hung Shao P'a Ya (红烧扒鸭)]

Ingredients:

- 1 duck (about 3 lbs.)
- 5 large mushrooms (soaked)
- 1 large slice of ginger
- 1 stalk of Chinese onion
- 2 teaspoonfuls of sesamum oil
- 3 teaspoonfuls of salt
- 3 teaspoonfuls of wine
- 3 teaspoonfuls of soya sauce
- 2 teaspoonfuls of sugar

Method:

Place the duck in a deep saucepan, and add the rest of the ingredients and sufficient water to cover the duck. Bring to the boil, then allow to simmer for 1½ hours. Serve whole, with gravy.

红烧扒鸭

材料：

一只鸭子（约三磅）

五只大冬菇，泡发

一大片生姜

一根青葱

二小匙麻油

三小匙细盐

三小匙米酒

三小匙酱油

二小匙白糖

做法：

把鸭子置在锅中，将以上的材料加入，需要水量和鸭子相平。煮沸后改用慢火，大约需一个半小时。食时整只装在大碗中，连原汁同餐。

FRIED DUCK LIVER
[Ch'ao Ya Kan (炒鸭肝)]

Ingredients:

8	duck livers (sliced)
2	mushrooms (soaked and sliced)
1	bamboo shoot (shredded)
1	small stalk of Chinese onion (sliced)
1	small slice of ginger
1/2	clove of garlic
a	dash of pepper
1/3	cup of lard
1/2	teaspoonful of salt
1/2	teaspoonful of sugar
1½	teaspoonfuls of cornflour
1	teaspoonful of sesamum oil
1	teaspoonful of wine
3	teaspoonfuls of soya sauce
1/3	cup of stock

Method:

To the liver add a dash of pepper, salt and a teaspoonful of cornflour, and mix well.

Heat the lard in a frying pan, fry the liver for 1 minute, remove and drain.

Using the same lard left in the pan, fry the garlic, ginger and onion, and brown a little.

Add the bamboo shoot, mushrooms, soya sauce, sugar, sesamum oil and stock, and allow to cook for 2 minutes.

Now replace the liver in the pan with the rest. Finally add the wine, and 1/2 teaspoonful of cornflour mixed with a little cold water.

炒鸭肝

材料：

八只鸭肝，切片

二只冬菇，泡发，切片

一只竹笋，切成薄片

一小根青葱，切碎

一小片生姜

半粒蒜头

少许胡椒粉

三分之一杯猪油

半小匙细盐

半小匙白糖

一又二分之一小匙玉米粉

一小匙麻油

一小匙米酒

三小匙酱油

三分之一杯原汁肉汤

做法：

鸭肝里加入胡椒粉、细盐和一小匙玉米粉，调顺之。

将猪油倒进炒锅中烧热，翻炒鸭肝一分钟即盛起。

将余下的猪油把蒜头、生姜、葱炒成金黄色，然后将竹笋、冬菇、酱油、白糖、麻油和原汁肉汤加入煮二分钟。

现在把那鸭肝倒进锅中，将酒、半小匙玉米粉和水调成糊加入锅中，翻动数下即盛起食之。

FRIED WILD DUCK
[Ch'ao Shui Ya P'ien (炒水鸭片)]

Ingredients:

2	ducks (breasts only, sliced)
2	large mushrooms (soaked and sliced)
1	bamboo shoot (sliced)
1	teaspoonful of sesamum oil
1	teaspoonful of sugar
a	dash of pepper
1/2	teaspoonful of cornflour
1	teaspoonful of salt
1	teaspoonful of wine
5	dessert-spoonfuls of cold water
1/3	cup of lard

Method:

To the duck meat add 1/2 teaspoonful of salt and a dash of pepper, then mix thoroughly with cornflour and water.

Put 1/8 cup of lard into a hot pan, and fry the mushrooms and bamboo shoot with 1/2 teaspoonful of salt and sugar. Turn out into a dish.

Add the rest of the lard to the pan, and fry the duck meat quickly for 1 minute. Drain off fat and add the wine and sesamum oil, together with mushrooms and bamboo shoot.

炒水鸭片

材料：

二只鸭子，取胸部肉切片

二只大冬菇，泡发，切片

一只竹笋，切片

一小匙麻油

一小匙白糖

少许胡椒粉

半小匙玉米粉

一小匙细盐

一小匙米酒

五中匙水

三分之一杯猪油

做法：

鸭肉片里加半小匙盐和少许胡椒粉，然后将玉米粉和水调和加入。

倒一点猪油在炒锅中烧热，将竹笋和冬菇炒之，同时加半小匙盐和糖，炒熟后盛在盘里。

把其余的猪油倒在炒锅里烧得很热，将鸭肉片用快速度翻炒一分钟。将油排出，再加酒和麻油，最后把冬菇、竹笋混合在鸭肉中略翻数下，盛起供餐。

MINCED PIGEON
[Kê Tzu Sung (鸽子松)]

Ingredients:

- 6   pigeons (only meat, coarsely minced)
- 4   small mushrooms (soaked and chopped)
- 10   water chestnuts (skin removed and finely chopped)
- 1   bamboo shoot (skin removed and finely diced)
- 1   stalk of celery (finely chopped)
- 1/4   lb. vermicelli (fried in deep fat)
- 1   stalk of Chinese onion (chopped)
- 1   teaspoonful of wine
- 1   teaspoonful of salt
- 2½   teaspoonfuls of cornflour
- 1/2   cup of meat sock
- 3   teaspoonfuls of soya sauce
- 1/2   teaspoonful of sugar
- 1/3   cup of lard

Method:

Mix 1/2 teaspoonful of salt and 1/2 teaspoonful of cornflour with the minced pigeon.

Heat the lard in a frying pan, and brown the onion. Then fry the minced pigeon, and add the soya sauce, sugar, celery, bamboo shoot, water chestnuts, mushrooms, meat stock, wine and the rest of the salt.

Mix 2 teaspoonfuls of cornflour with a little water, and add to the contents of the pan.

Now break up the fried vermicelli with the fingers, place on a dish, and pour the pigeon mixture over it for serving.

鸽子松

材料：

六只鸽子的肉，粗粗切碎

四只小冬菇，泡发，切碎

十只荸荠，去皮，切碎

一只竹笋，去皮，切碎

一大根芹菜，切碎

三两面粉条，用油炸黄

一根青葱，切碎

一小匙米酒

一小匙细盐

二又二分之一小匙玉米粉

半杯原汁肉汤

三小匙酱油

半小匙白糖

三分之一杯猪油

做法：

用半小匙细盐和半小匙玉米粉调在鸽子肉中。

将油置在炒锅里，把葱炒成金黄色。然后将鸽子肉和酱油、糖、芹菜、竹笋、荸荠、冬菇、原汁肉汤、酒及余下的细盐等全部加入炒之。

余下的二小匙玉米粉用一点水将它调和，加入在以上材料里。

现在你可将炸好的面条用手弄开，装在大盘里，最后把那作料倒在上头食之。

MEAT DISHES
[Jou Lei (肉类)]

Pages

135—136	Fried Beef Fillet
138	Egg and Beef Omelets
140—141	Stuffed Green Peppers
144	Brown Sauce Beef
146	Stewed Shin of Beef
148—149	Sweet and Sour Pork (Boneless)
152	Red Sauce Rump of Pork
154—155	Stewed Meatballs with Shantung Cabbage
158	Stewed Rice Flour Pork
160	Roast Shaslick Pork
162	Stewed Pork with Bamboo Shoots

肉　类

137　　　　　炒牛里脊

139　　　　　煎鸡蛋饺

142—143　　牛肉酿辣椒

145　　　　　锅烧牛肉

147　　　　　红烧牛肉

150—151　　糖醋排骨

153　　　　　扒肘子

156—157　　红烧狮子头

159　　　　　米粉肉

161　　　　　叉烧肉

163　　　　　红焖五花猪肉

Kiangsi Porcelain
江西瓷器

FRIED BEEF FILLET
[Ch'ao Niu Li Chi (炒牛里脊)]

Ingredients:

1½	lbs. beef fillet (sliced)
1	egg white
1	small cauliflower (cooked and sliced)
1	large onion (sliced)
1	green pepper (sliced)
1/2	cup of lard
2	teaspoonfuls of salt
a	dash of pepper
2	teaspoonfuls of wine
2	teaspoonfuls of soya sauce
2	teaspoonfuls of cornflour
20	teaspoonfuls of cold water

Method:

Mix the beef well with the pepper, cornflour, salt and 1 teaspoonful of wine. Add the egg white and water to make the pieces adhere.

Heat the lard in a frying pan, and when boiling hot fry the beef for 1 minute. Remove and drain, leaving a little lard in the pan. Fry the onion to a light brown, and then add cauliflower, green pepper and soya sauce.

Now replace the beef in the pan, add 1 teaspoonful of wine, stir well with the ingredients for about 2 minutes, and serve.

炒牛里脊

材料：

一又二分之一磅牛里脊肉，切薄片

一只鸡蛋的白

一个小花菜，煮熟，切片

一个大洋葱，切丝

一只青椒，切丝

半杯猪油

二小匙细盐

少许胡椒粉

二小匙米酒

二小匙酱油

二小匙玉米粉

廿小匙冷水

做法：

将牛肉和胡椒粉、玉米粉、细盐及一小匙酒等混合调顺，再加入蛋白和水依附之。

把油倒在炒锅里烧热，将牛肉炒一分钟，将它盛起，留余下的油炒洋葱之用，炒到洋葱发黄时，然后加入花菜、青椒和酱油。

现在再把牛肉倒回锅里，加一小匙酒，与其他所有的作料再翻炒二分钟，即供食。

EGG AND BEEF OMELETS
[Chien Chi Tan Chiao (煎鸡蛋饺)]

Ingredients:

- 8 whole eggs (lightly beaten)
- 2 teaspoonfuls of soya sauce
- 1/2 teaspoonful of sugar
- 1/2 onion (cut up very finely)
- 1/3 cup of lard
- 1 lb. lean beef (minced) ⎫
- 1/2 teaspoonful of salt ⎬ mix well together
- 1/2 teaspoonful of cornflour ⎪
- 1 teaspoonful of wine ⎭

Method:

Brown the onion in 1/4 cup of lard, then put in the meat mixture and fry. Add the soya sauce and sugar. When the meat is nicely browned, turn it out onto a dish.

Add a little more lard. When very hot, pour in about 1 teaspoonful of beaten egg. Allow it to set slightly, then place a small quantity of the meat mixture on it, fold over, and fry. Repeat this until all the eggs are used up. This is enough for about 20 individual omelets.

煎鸡蛋饺

材料：

八只鸡蛋，搅之　　　　　一磅牛肉，搅细

二小匙酱油　　　　　　　半小匙细盐

半小匙白糖　　　　　　　半小匙玉米粉

半个洋葱，切得很细　　　一小匙米酒

三分之一杯猪油　　　　　将以上四种材料混合之。

做法：

用四分之一杯的油将洋葱炒成褐色，然后将牛肉放入炒之，加入酱油和糖，熟后将它盛出。

再加入一点油在炒锅里，够热时，盛一小匙蛋液在锅中煎成圆薄片，让停一会，将少量肉置在中央，包起像饺子似的，炸之。这些材料做完后，共有廿只蛋饺子。

STUFFED GREEN PEPPERS
[Niu Jou Jang La Chiao (牛肉釀辣椒)]

Ingredients:

1	lb. green peppers (seeds removed)
1½	lbs. beef (minced)
1	piece of Chinese onion (chopped finely)
1	small slice of ginger (chopped finely)
2	teaspoonfuls of wine
2	teaspoonfuls of salt
4	teaspoonfuls of soya sauce
1	teaspoonful of cornflour
1/4	cup of lard
1	cup of stock
1	heaped teaspoonful of sugar
2	teaspoonfuls of sesamum oil
20	teaspoonfuls of water

Method:

Mix the beef with onion, ginger, wine, salt, cornflour, 2 teaspoonfuls of soya sauce, sesamum oil and cold water well. Stuff the peppers with the mixture.

Sauté in 1/4 cup of lard in a hot frying pan, add 2 teaspoonfuls of soya sauce, sugar and 1 cup of stock, turn into a deep saucepan, bring to a boil, and simmer for about 30 minutes.

牛肉酿辣椒

材料：

　　一磅青辣椒，除去籽

　　一又二分之一磅牛肉，搅细

　　一根青葱，切得很细

　　一小片生姜，切得很细

　　二小匙米酒

　　二小匙细盐

　　四小匙酱油

　　一小匙玉米粉

　　四分之一杯猪油

　　一杯原汁肉汤

　　一小满匙白糖

　　二小匙麻油

　　廿小匙冷水

做法：

先将牛肉和葱、姜、酒、盐、玉米粉、二小匙酱油、麻油及水等混合之，然后再将它装在青椒内。

将青椒放入热炒锅用四分之一杯猪油略煎，加入二小匙酱油、白糖及一杯原汁肉汤。转置在深底锅子里，将青椒排列在里面，用文火煮三十分钟之久，就可供食。

BROWN SAUCE BEEF
[Kuo Shao Niu Jou (锅烧牛肉)]

Ingredients:

1½	lbs. beef (ribs)
1	clove of garlic
1	piece of ginger
1	stalk of Chinese onion
1	grain of star aniseed
2	teaspoonfuls of sesamum oil
2	teaspoonfuls of salt
1	teaspoonful of sugar
6	teaspoonfuls of cornflour
2	teaspoonfuls of soya sauce
1	egg

Method:

To the beef in a deep saucepan, add all the ingredients (except the cornflour and egg) and enough water to cover it. Bring to the boil and then simmer for 3 hours. Take the beef out.

Make a paste with the cornflour, egg and the gravy from the beef, and cover the beef with it. Then fry the beef in deep fat until golden brown. Serve in slices.

锅烧牛肉

材料：

一又二分之一磅牛肉（肋部）

一粒蒜头

一小块生姜

一根青葱

一个八角

二小匙麻油

二小匙细盐

一小匙白糖

六小匙玉米粉

二小匙酱油

一只鸡蛋

做法：

将牛肉及所有的材料（除玉米粉及鸡蛋）置在锅中，用足够的水量盖着牛肉。煮沸之后将火调低煮至三个小时，然后把牛肉盛起。

将玉米粉和蛋调顺加入牛肉原汁汤里，覆于牛肉上。最后将牛肉用油炸成金黄色。食时切成片。

STEWED SHIN OF BEEF
[Hung Shao Niu Jou (红烧牛肉)]

Ingredients:

- 2 lbs. shin of beef
- a dash of pepper
- 1 large slice of ginger
- 1 clove of garlic
- 1 stalk of Chinese onion
- 10 teaspoonfuls soya sauce
- 1 teaspoonful of sesamum oil
- 2 teaspoonfuls of salt
- 2 teaspoonfuls of sugar

Method:

Sauté the above ingredients together for a few minutes. Turn into a deep saucepan, and add sufficient water to cover the meat. Bring to the boil. Simmer for 3 hours.

Slice and serve with the gravy.

红烧牛肉

材料：

二磅牛腿肉

少许胡椒粉

一大片生姜

一粒蒜头

一根青葱

十小匙酱油

一小匙麻油

二小匙细盐

二小匙白糖

做法：

首先将全部的材料炒一下，再把它倒进汤锅里，需充分的水量盖着牛肉。煮沸之后，调低火候煮三小时。

食时切成片，同原汁汤食之。

SWEET AND SOUR PORK (BONELESS)
[T'ang Ts'u P'ai Ku (糖醋排骨)]

Main Ingredients:

- 2 lbs. loin of pork (boned)
- 1 teaspoonful of salt
- 1 teaspoonful of Chinese wine
- 3 teaspoonfuls of soya sauce
- 6 heaped teaspoonfuls of cornflour

Sauce Ingredients:

- 6 pieces of red haws
- 1/2 cup of Chinese sweet and sour mixed pickles (or any other kind)
- 1 small piece of Chinese onion (finely chopped)
- 1/8 cup of lard
- 1/4 cup of vinegar
- 6 teaspoonfuls of sugar
- 1/2 cup of water
- 1/4 clove of garlic (finely chopped)
- 3 teaspoonfuls of soya sauce
- 1½ teaspoonfuls of cornflour

Method:

Cut the pork into fairly large squares and mix well with the above ingredients. Fry in deep fat until crisp and a golden brown. Drain and turn out on a plate.

Heat the lard in a pan, and brown the onion and garlic a little. Pour in the above ingredients (except the cornflour, water and pork), adding the cornflour previously mixed with a little water, slowly, and stirring, for 1 minute.

Then pour this sauce over the pork. Serve very hot while the pork is crisp. The above quantity is sufficient for 6 to 8 persons.

糖醋排骨

主菜的材料：

二磅排骨肉，去骨，切成大小相当的方块

一小匙细盐

一小匙米酒

三小匙酱油

六小满匙玉米粉

以上的材料全部混合之。

调味汁的材料：

六块山楂饼　　　　　　六小匙白糖

半杯中国的泡菜　　　　半杯水

一小根青葱，切碎　　　四分之一粒蒜头，切碎

八分之一杯猪油　　　　三小匙酱油

四分之一杯醋　　　　　一又二分之一小匙玉米粉

做法：

用充分的油量将肉炸成金黄色，然后将它盛出，置在盘子里备用。

将油倒在锅里烧热，把葱、蒜头略炒数下，再将全部其他材料（除玉米粉、水和肉以外）加入，用一点水把玉米粉调成糊徐徐加进，煮一分钟。

将做好的汤汁浇于炸好的肉上，趁肉热而脆时供食。这可供六至八位食之。

RED SAUCE RUMP OF PORK
[P'a Chou Tzu (扒肘子)]

Ingredients:

- 2   lbs. pork rump
- 1   large piece of ginger
- 1/2 stalk of Chinese onion
- 1/4 cup of lard
- 3   teaspoonfuls of soya sauce
- 2   teaspoonfuls of salt
- 2   teaspoonfuls of sugar
- 1   teaspoonful of sesamum oil

Method:

Heat the lard in a frying pan. Fry the rump with the ginger and onion. Then add the soya sauce, frying for about 5 minutes.

Now turn out into a deep saucepan, then add the salt, sesamum oil, sugar and sufficient water to cover. Bring to the boil, and allow to simmer for 2 hours. Serve with gravy.

扒肘子

材料：

二磅猪腿肉

一大片生姜

半根青葱

四分之一杯猪油

三小匙酱油

二小匙细盐

二小匙白糖

一小匙麻油

做法：

将油倒在炒锅中烧热，把肉、葱和姜炒几下。然后加入酱油，炒约五分钟。

现在将肉盛出，转置在深底的汤锅中，再加入细盐、麻油和白糖，用充分的水量盖住肉。煮沸后改用慢火候煮二小时，食时同原汁汤供餐。

STEWED MEATBALLS WITH SHANTUNG CABBAGE
[Hung Shao Shih Tzu T'ou (红烧狮子头)]

Ingredients for Meatballs:

- 1 lb. beef or pork (finely minced)
- 1 small slice of ginger (chopped)
- 1 small stalk of Chinese onion (chopped)
- 1 teaspoonful of wine
- 2 teaspoonfuls of cornflour
- 1 teaspoonful of sesamum oil
- 3 teaspoonfuls of water
- 1/2 teaspoonful of salt
- 1/3 cup of lard

Ingredients for Cabbage:

- 2 cabbages (coarsely chopped)
- 6 teaspoonfuls of soya sauce
- 1 cup of stock
- 2 teaspoonfuls of sugar
- 1/2 teaspoonful of salt

Method:

Mix the above ingredients (except the lard) thoroughly into a stiff mass, then make into 4 large meatballs.

Heat the lard in a pan, and sauté the meatballs until golden brown on both sides. Leave in a dish.

Sauté the cabbage in the fat left in the pan for 2 to 3 minutes, after adding the soya sauce, sugar, salt and stock. Turn out into a deep pan, place the meatballs on top of the cabbage, and simmer for 1 hour.

红烧狮子头

肉丸的材料：

　　一磅牛肉或猪肉，搅细

　　一小片生姜，切碎

　　一小根青葱，切碎

　　一小匙米酒

　　二小匙玉米粉

　　一小匙麻油

　　三小匙水

　　半小匙细盐

　　三分之一杯猪油

配菜的材料：

　　二个白菜，叶不要切得太碎

　　六小匙酱油

　　一杯原汁肉汤

　　二小匙白糖

　　半小匙细盐

做法：

以上所有的肉丸材料（除猪油外）全部混合调顺，搅得很集密之后，将它做成四个大肉圆。

将猪油倒在炒锅中烧热，把大肉圆两面煎成金黄色，盛出在盘上备用。

把炒锅里剩下的油烧热，放入白菜，然后加入酱油、白糖、细盐和原汁肉汤炒二至三分钟，将之转移到深底的汤锅中，将肉圆放置在白菜的上头，用慢火候煮上一小时，即可供食。

STEWED RICE FLOUR PORK
[Mi Fên Jou(米粉肉)]

Ingredients:

1	cup of uncooked rice
1½	lbs. streaky pork
1	piece of ginger (finely chopped)
1	stalk of Chinese onion (finely chopped)
1/2	cup of water
2	teaspoonfuls of sesamum oil
4	teaspoonfuls of soya sauce
1/2	teaspoonful of sugar
1/2	teaspoonful of salt

Method:

Fry the rice in a hot dry pan, then pulverise it. Put the powder into a bowl, then add the ginger, onion, salt, sugar, soya sauce, sesamum oil and water to make into a paste.

Boil the pork for 1 hour, then cut it up into thin slices. These are set in a deep bowl side by side, with a layer of the paste intervening, and with the skin surface downwards. They are then steamed for 2½ hours, with a thick slice of ginger and a large piece of Chinese onion. The whole is turned out onto a dish, the lower side up, for serving.

米粉肉

材料：

一杯生米

一又二分之一磅五花肉

一片生姜，部分切碎

一根青葱，部分切碎

半杯水

二小匙麻油

四小匙酱油

半小匙白糖

半小匙细盐

做法：

首先将米放在热锅中炒之，然后把它磨碎。把米粉置在碗中，加入生姜、青葱、细盐、白糖、酱油、麻油和水，调成糊。

把肉煮上一个小时，然后将它切成薄片。将肉片并排置在深碗中，带皮的一面朝下，中间糊上米糊。最后放一片厚生姜和一大根青葱在上面，将它蒸二个半小时。好了之后，把整碗肉翻在盘子上食之。

ROAST SHASLICK PORK
[Ch'a Shao Jou (叉烧肉)]

Ingredients:

- 1½ lbs. loin pork
- A dash of pepper
- 1 small stalk of Chinese onion
- 1 slice of ginger
- 1/4 teaspoonful of thick sauce (chiang)
- 1/2 teaspoonful of sugar
- 5 teaspoonfuls of soya sauce
- 1 teaspoonful of wine
- 1/2 teaspoonful of salt

Method:

Remove the bone from the pork. Cut it into strips about 4 inches long. Mix all the ingredients with the pork thoroughly and allow to soak for 1/2 hour.

Roast in a hot oven for 10 minutes, then turn the meat over and roast for another 5 to 10 minutes.

Slice before serving.

叉烧肉

材料：

一又二分之一磅排骨肉

少许胡椒粉

一小根青葱

一片生姜

四分之一小匙豆酱

半小匙白糖

五小匙酱油

一小匙米酒

半小匙细盐

做法：

将已去骨的排骨肉切成四英寸长条，和以上全部其他材料混合，泡上半个小时。

将它烤十分钟，然后把它翻一个身再烤五至十分钟。

吃的时候切成片供食。

STEWED PORK WITH BAMBOO SHOOTS
[Hung Mên Wu Hua Chu Jou(红焖五花猪肉)]

Ingredients:

- 1/2 lb. pork (boiled 1 hour, then cut into squares)
- 2 bamboo shoots (cut into squares)
- 2 mushrooms (soaked and cut into squares)
- 1 small stalk of Chinese onion
- 1/2 clove of garlic
- 1 slice of ginger
- 2 teaspoonfuls of sugar
- 2 cups of water
- 10 teaspoonfuls of soya sauce
- 1 teaspoonful of salt
- 1/4 cup of lard

Method:

Heat the lard in a frying pan, then fry the ginger, garlic and onion for 1 minute.

Add mushrooms, bamboo shoots and pork, then the salt, sugar, soya sauce and cold water, and cook for about 3 minutes.

Turn out into a deep saucepan, and simmer for 1½ hours. Serve with gravy.

红焖五花猪肉

材料：

半磅五花肉，约煮一小时，然后切丁

二只竹笋，切丁

二只冬菇，泡发，切成四开

一小根青葱

半粒蒜头

一片生姜

二小匙白糖

二杯水

十小匙酱油

一小匙细盐

四分之一杯猪油

做法：

将猪油置在炒锅中烧热，将姜、蒜、葱炒一分钟。

加入冬菇、竹笋和肉，然后再加细盐、白糖、酱油和水，约煮三分钟。

把它转至深底的汤锅里，用慢火煮一个半小时。吃时和原汁汤同食之。

SEAFOOD DISHES
[Hê Hai Hsien Lei(河海鲜类)]

Pages

167	Brown Sauce Mandarin Fish
170—171	Fried Fish with Sour Sauce
174—175	Smoked Fish
178—179	Fried Minced Fish
182	White Sauce Fish (Sour)
184—185	Fried Fish Cakes
188—189	Braised Shrimps
192	Shrimp Cakes
194	Fried Shrimp Balls
196	Stuffed Clams
198	Crab Omelet
200	Crab Fat with Green Vegetables (Green Jade and Red Coral)

河海鲜类

168—169	红烧鳜鱼
172—173	炸松鼠鱼
176—177	熏鱼
180—181	炒鱼松
183	五柳鳜鱼
186—187	煎鱼饼
190—191	烩虾仁
193	煎虾饼
195	炸虾球
197	酿海蛤蜊
199	芙蓉蟹肉
201	碧玉珊瑚

Wine Pot and Cups
酒壶和杯子

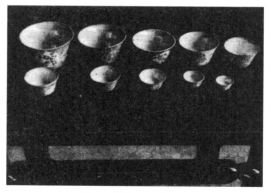

Ten Wine Cups
十个酒杯

BROWN SAUCE MANDARIN FISH
[Hung Shao Kuei Yü (红烧鳜鱼)]

Ingredients:

1	mandarin fish (about 2 lbs.)
1/4	lb. fat pork (cut into squares)
6	mushrooms (soaked)
1/2	bamboo shoot (sliced)
1	slice of ginger
1	small stalk of Chinese onion
1	large clove of garlic
1/3	cup of lard
7	teaspoonfuls of soya sauce
1	teaspoonful of salt
2	teaspoonfuls of wine
1	cup of meat stock
1/2	teaspoonful of sugar
1	piece of star aniseed

Method:

Heat the lard in a frying pan, and sauté the fish, which has been previously scored, for 5 minutes until it is brown on both sides.

Add in the ginger, onion, garlic, mushrooms, bamboo shoot, soya sauce, salt, wine, fat pork and meat stock. Cook for a while, and finally add the sugar and aniseed. Bring to the boil, then simmer for 1/2 hour with a cover on the frying pan. Serve the fish whole.

红烧鳜鱼

材料：

一条鳜鱼（约二磅）

四分之一磅肥猪肉，切丁

六只冬菇，泡发

半只竹笋，切片

一片生姜

一小根青葱

一大粒蒜头

三分之一杯猪油

七小匙酱油

一小匙细盐

二小匙米酒

一杯原汁肉汤

半小匙白糖

一粒八角

做法：

　　把猪油倒在炒锅中烧热，将事先已切好花刀的鱼两面煎成金黄色，大约需五分钟。

　　再加入生姜、青葱、蒜头、冬菇、竹笋、酱油、细盐、米酒、肥肉和原汁肉汤，将它煮一会，再加进白糖和八角。沸腾后把锅的盖子盖上，用慢火候煮烧半个小时。好了之后，整条鱼上桌供食。

FRIED FISH WITH SOUR SAUCE
[Cha Sung Shu Yü (炸松鼠鱼)]

Main Ingredients:

- 1   mandarin fish (about 3 lbs.)
- 6   teaspoonfuls of cornflour
- 1   teaspoonful of salt
- 2   teaspoonfuls of wine

Sauce Ingredients:

- 1   small stalk of Chinese onion
- 1   small slice of ginger
- 3   teaspoonfuls of sugar
- 2   teaspoonfuls of cornflour
- 2   teaspoonfuls of soya sauce
- 1/4 cup of vinegar
- 1/4 cup of lard
- 1/4 lb. pork (sliced)

Method:

Bone the fish and score the outside. Add the wine and salt. Rub thoroughly with 6 teaspoonfuls of cornflour and fry the fish until golden brown and crisp.

Now make the sauce as follows:

Heat the lard in a frying pan. Fry the ginger and onion a little, then add 2 teaspoonfuls of soya sauce, the vinegar, sugar and sliced pork.

Mix 2 teaspoonfuls of cornflour with a little water, and add slowly, stirring all the time, then pour the sauce on the fish and serve.

炸松鼠鱼

主菜的材料：

    一条鳜鱼（约三磅）

    六小匙玉米粉

    一小匙细盐

    二小匙米酒

调味汁的材料：

    一小根青葱

    一小片生姜

    三小匙白糖

    二小匙玉米粉

    二小匙酱油

    四分之一杯米醋

    四分之一杯猪油

    四分之一磅猪肉，切片

做法：

首先将鱼洗涤去骨，两面划上几刀，抹一点酒和细盐，再擦上一些玉米粉（约六小匙），将鱼炸成金黄色。

现在可以做那些调味料，如下：

将猪油置在炒锅里烧热。把姜、葱略炒一下，然后加入二小匙酱油、米醋、白糖和猪肉片。

用一点水将二小匙玉米粉调和，慢慢地加入的同时不停地拌之。这些调味料做好之后，倒在鱼的上头供食。

SMOKED FISH
[Hsün Yü (熏鱼)]

Main Ingredients:

- 1½ lbs. yellow fish or any other fish (cut into large slices)
- 1 large slice of ginger
- 1 large stalk of Chinese onion
- 1 clove of garlic
- a dash of pepper
- 8 teaspoonfuls of soya sauce
- 1 teaspoonful of sesamum oil
- 3 teaspoonfuls of Chinese wine
- 2 teaspoonfuls of sugar

Sauce Ingredients:

- 2 teaspoonfuls of soya sauce
- 1/2 teaspoonful of salt
- 1 teaspoonful of sugar
- 1/3 cup of stock

} mix together

Method:

Soak the fish with the other main ingredients for 1 hour. Fry the fish, piece by piece, in deep fat, until golden brown.

Fry the fish again in 1/4 cup of lard, then add the above sauce, and cook until the sauce has evaporated. The fish can be served either hot or cold.

熏　鱼

主菜的材料：

　　一又二分之一磅黄鱼或其他的鱼，切成大片

　　一大片生姜

　　一大根青葱

　　一粒蒜头

　　少许胡椒粉

　　八小匙酱油

　　一小匙麻油

　　三小匙米酒

　　二小匙白糖

调味汁的材料：

| 二小匙酱油 | 一小匙白糖 |
| 半小匙细盐 | 三分之一杯原汁肉汤 |

以上的作料混合在一起。

做法：

  首先将切好的鱼片和以上其他主菜材料混合泡至一小时，然后将鱼片逐一徐徐地放入热油中炸成金黄色。

  用四分之一杯猪油再将鱼片炸一下，然后加入全部调味材料煮到快没有水分时为止。这道菜可吃热的或冷的。

FRIED MINCED FISH
[Ch'ao Yü Sung (炒鱼松)]

Ingredients:

- 1/2 lb. fish (minced)
- 1/2 lb. turnips (shredded)
- 1 stalk of Chinese onion (finely chopped)
- 1 thin slice of ginger (finely chopped)
- 1 egg white
- 1 heaped teaspoonful of cornflour
- 3/4 cup of lard
- 1 teaspoonful of salt
- 4 teaspoonfuls of wine
- 1 teaspoonful of soya sauce
- 1/2 teaspoonful of sugar
- 1/2 cup of peanuts (ground)
- a dash of pepper

Method:

Mix the fish well with a dash of pepper, 1/2 teaspoonful of salt, the cornflour, 3 teaspoonfuls of wine and an egg white.

Heat 1/2 cup of the lard in a frying pan, and fry the turnips for 3 or 4 minutes, adding 1/2 teaspoonful of salt. Then add 1 teaspoonful of soya sauce, 1/2 teaspoonful of sugar, and fry for another few minutes. Turn out into a dish.

Put the rest of the lard into the pan, brown the ginger and onion a little, then fry the fish mixture together until it becomes lumpy. Add 1 teaspoonful of wine and cook for 1 minute. Pour this out on top of the turnips, garnish with the ground peanuts and serve.

炒鱼松

材料:

半磅鱼,切碎

半磅白萝卜,切片

一根青葱,切碎

一小片生姜,切碎

一只鸡蛋的白

一小满匙玉米粉

四分之三杯猪油

一小匙细盐

四小匙酒

一小匙酱油

半小匙白糖

半杯花生米,磨碎

少许胡椒粉

做法：

将鱼与胡椒粉、半小匙细盐、玉米粉、三小匙酒和蛋白混合之。

把炒锅烧热，倒进半杯猪油，将白萝卜炒三至四分钟，加半小匙细盐，后加入一小匙酱油和半小匙白糖再炒数分钟，将它盛在盘里。

将其余的猪油倒在锅中，把生姜、葱炒成黄色，然后把鱼加入混合炒之。再加入一小匙酒炒一分钟。将它倒在萝卜上面，用花生粉置在上头供食。

WHITE SAUCE FISH (SOUR)
[Wu Liu Kuei Yü (五柳鳜鱼)]

Ingredients:

1	whole mandarin fish (scored)
1	small stalk of Chinese onion
1	small piece of ginger
1/2	clove of garlic
2	teaspoonfuls of sweet and sour pickles (Chinese or foreign)
1/3	cup of vinegar (white)
2	teaspoonfuls of cornflour
1/2	teaspoonful of salt
6	teaspoonfuls of sugar
1/2	cup of fish stock
1/4	cup of lard

Method:

Put the fish in a pan of boiling water and boil for 15 minutes. Leave it in a dish.

Then make the sauce. Heat half of the lard in a pan, then fry the ginger, onion and garlic together with the vinegar, stock, sugar, salt and pickles. Now mix the cornflour with a little cold water, and add it to the contents of the pan. Finally add the other half of the lard. Pour this sauce over the fish and serve.

五柳鳜鱼

材料：

一条鳜鱼，划线

一小根青葱

一小片生姜

半粒蒜头

二小匙酱黄瓜（中国的或外国的）

三分之一杯白醋

二小匙玉米粉

半小匙细盐

六小匙白糖

半杯原汁鱼汤

四分之一杯猪油

做法：

首先把鱼放在沸水里煮十五分钟，盛在盘里。

接着做调味料：倒一半猪油在锅中，将生姜、葱、蒜头一起炒，再加醋、原汁鱼汤、白糖、细盐和酱瓜。现在将玉米粉和少许水调糊，慢慢地加入。再将余下一半猪油加进。吃时把这调味料倒在鱼上头供食。

FRIED FISH CAKES
[Chien Yü Ping (煎鱼饼)]

Ingredients:

- 1/4 lb. fish (minced)
- 1/4 cup of dry shrimps (pounded)
- 1 tin of bamboo shoot (coarsely sliced)
- 8 small mushrooms (soaked)
- 1 slice of ham (cut into big pieces)
- 1/4 cup of lard
- 1/4 cup of stock
- 1 teaspoonful of salt
- 1/2 teaspoonful of sugar
- 1 teaspoonful of wine
- 1 teaspoonful of cornflour
- 2 teaspoonfuls of soya sauce
- 1 teaspoonful of sesamum oil

Method:

Mix the fish well with 1/2 teaspoonful of salt, 1/2 teaspoonful of cornflour, 1/4 cup of dry shrimps and 1/4 cup of stock. Shape into small cakes. Heat the lard in a pan, and sauté the cakes on both sides. Turn out onto a dish.

Using the same fat, fry the bamboo shoot, mushrooms and ham for 1 minute. Add the sugar, soya sauce, wine and the rest of the salt. Replace the fish cakes. Mix 1/2 teaspoonful of cornflour with a little water, and add slowly. Lastly add the sesamum oil to flavour.

煎鱼饼

材料：

四分之一磅鱼，切碎

四分之一杯虾米，捣碎

一罐竹笋，不要切得太碎

八只小冬菇，泡发

一片火腿，切成大片

四分之一杯猪油

四分之一杯原汁肉汤

一小匙细盐

半小匙白糖

一小匙米酒

一小匙玉米粉

二小匙酱油

一小匙麻油

做法:

将半小匙盐、半小匙玉米粉、四分之一杯虾米和四分之一杯原汁肉汤与鱼混合搅顺，做成饼的模样。把炒锅烧热，加点油将鱼饼两面煎成金黄色，盛在盘子里备用。

用那余在锅里的油将竹笋、冬菇和火腿略炒一分钟。再加入白糖、酱油、酒和剩下的细盐。之后把鱼饼置在锅中。将半小匙玉米粉用一点水调糊，徐徐地加入在锅里。倒点麻油供餐。

BRAISED SHRIMPS
[Hui Hsia Jên (烩虾仁)]

Ingredients:

1/2	lb. shrimps (shelled)
6	water chestnuts (cooked and cut into small squares)
4	mushrooms (cooked and cut into small squares)
1/2	cup of peas (tinned)
2	slices of cooked Chinese ham (cut into small squares)
1	tiny piece of ginger (chopped finely)
1	tiny piece of Chinese onion (chopped finely)
1	teaspoonful of salt
2	teaspoonfuls of cornflour
2	teaspoonfuls of soya sauce
2	teaspoonfuls of wine
1	teaspoonful of sesamum oil
1/3	cup of lard
1	cup of stock

Method:

Heat the lard in a frying pan, and fry the shrimps a little together with the ginger, onion, soya sauce, wine, stock and salt.

Add the water chestnuts, mushrooms, peas and ham. Now mix the cornflour with a little cold water and add it to the mixture.

Finally add the sesamum oil. Fry altogether for 3 minutes.

烩虾仁

材料：

半磅虾仁

六只荸荠，煮熟，切成小丁

四只冬菇，泡发，煮熟，切小丁

半杯罐装青豆

二片金华熟火腿，切小丁

一小片生姜，切碎

一小根青葱，切碎

一小匙细盐

二小匙玉米粉

二小匙酱油

二小匙酒

一小匙麻油

三分之一杯猪油

一杯原汁肉汤

做法：

将猪油倒在炒锅中烧热，把虾仁略炒一下，同时加姜、葱、酱油、酒、原汁肉汤和细盐。

再加入荸荠、冬菇、青豆和火腿。现在用一点水把玉米粉调糊，徐徐地加入。

最后加入麻油，一起炒三分钟。

SHRIMP CAKES
[Chien Hsia Ping (煎虾饼)]

Ingredients:

- 1½ lbs. shrimps (shelled and pounded)
- 10 water chestnuts (skin removed and pounded)
- 1 whole egg
- a dash of pepper
- 1 slice of ginger (minced)
- 1 small piece of Chinese onion (minced)
- 1½ teaspoonfuls of salt
- 1 heaped teaspoonful of cornflour
- 1/3 cup of cold water or stock
- 1 Chinese cabbage (coarsely cut and part boiled)

Method:

Mix all the above ingredients together with chopsticks (except cabbage). Shape into small flat cakes, and sauté in hot lard until a nice golden colour on both sides.

Using the same lard fry the cabbage with 1 teaspoonful of salt for 2 minutes. Replace the shrimp cakes and fry altogether for another minute, then serve.

煎虾饼

材料：

一又二分之一磅虾仁，捣碎

十只荸荠，去皮，捣碎

一只鸡蛋

少许胡椒粉

一片生姜，切碎

一小根青葱，切碎

一又二分之一小匙细盐

一小满匙玉米粉

三分之一杯水或原汁肉汤

一个白菜，切粗点煮之

做法：

将以上除白菜外全部的材料用筷子搅顺，然后做成饼形模样，用点油把虾饼双面煎成金黄色。

用余在锅里的油将已煮过的白菜加一小匙盐炒二分钟。最后把虾饼加置在白菜中再煮烧一分钟，然后供餐。

FRIED SHRIMP BALLS
[Cha Hsia Ch'iu (炸虾球)]

Ingredients:

- 1¼ lbs. shrimps (shelled and minced)
- 6 water chestnuts (chopped finely)
- 1 egg white
- 1 slice of ginger (chopped finely)
- a little Chinese onion (chopped finely)
- 1 teaspoonful of salt
- 1 teaspoonful of cornflour
- 3 teaspoonfuls of wine
- 1 teaspoonful of sesamum oil
- a dash of pepper

Method:

Mix the above ingredients well, shape into balls, and fry them in deep fat until golden brown.

炸虾球

材料：

一又四分之一磅虾仁，切碎

六只荸荠，切碎

一只鸡蛋的白

一片生姜，切碎

一些青葱，切碎

一小匙细盐

一小匙玉米粉

三小匙米酒

一小匙麻油

少许胡椒粉

做法：

将以上所有的材料搅得很顺和，然后做成虾球，用多点油将虾球炸成金黄色盛出，供食。

STUFFED CLAMS
[Jang Hai Kê Li (酿海蛤蜊)]

Ingredients:

- 2½ lbs. clams (with shell)
- 1/4 lb. fat pork or beef (minced)
- 1 slice of ginger (finely chooped)
- 1 tiny piece of Chinese onion (finely chopped)
- 3 teaspoonfuls of wine
- 1/2 teaspoonful of salt
- 1/2 teaspoonful of sugar
- 1 teaspoonful of soya sauce
- 1/4 cup of stock
- 1/8 cup of lard

Method:

Pour boiling water over the clams and remove them from their shells. Mince the clams and the pork (or beef) together. Add to the mince the ginger, onion, wine, salt, sugar and soya sauce, and mix well.

Stuff the mixture into the empty shells. Place these in a baking dish containing the lard and stock, and bake in a hot oven for 10 minutes.

酿海蛤蜊

材料：

二又二分之一磅蛤蜊（带壳）

四分之一磅肥猪肉或牛肉，搅碎

一片生姜，切碎

一小根青葱，切碎

三小匙米酒

半小匙细盐

半小匙白糖

一小匙酱油

四分之一杯原汁肉汤

八分之一杯猪油

做法：

用开水将蛤蜊泡熟去壳，把蛤蜊肉切碎，和猪肉（或牛肉）混合之。加生姜、葱、酒、细盐、白糖和酱油，搅调得很顺和。

然后将混合好的材料装置在蛤蜊的壳中，加入油和原汁肉汤，在烤盘里一起烤十分钟，供食。

CRAB OMELET
[Fu Jung Hsieh Jou (芙蓉蟹肉)]

Ingredients:

- 2   crabs (cooked and shelled)
- 4   eggs
- 1   teaspoonful of salt
- 1/3 cup of lard
- 1   small stalk of Chinese onion (chopped finely)

Method:

Heat the lard in a frying pan, and brown the onion.

Beat the eggs lightly in a bowl, add the crab meat and salt, and fry together for 2 minutes. Serve hot.

芙蓉蟹肉

材料：

二只蟹，煮熟，去壳

四只鸡蛋

一小匙细盐

三分之一杯猪油

一小根青葱，切碎

做法：

将油倒在炒锅中烧热，把葱炒成黄色。

把鸡蛋轻轻地打在碗里，加入蟹肉和盐，一起炒二分钟，随即供餐。这道菜要吃热的。

CRAB FAT WITH GREEN VEGETABLES
(Green Jade and Red Coral)

[Pi Yü Shan Hu(碧玉珊瑚)]

Ingredients:

1/2	lb. crab fat (cooked)
3½	lbs. mustard greens (only young stems) or any other green vegetables
1	slice of ginger
1/3	cup of lard
1	teaspoonful of salt
3	teaspoonfuls of wine
4	teaspoonfuls of soya sauce
1	teaspoonful of sugar

Method:

Heat the lard in the frying pan. Fry the ginger a little, then remove the ginger. Add the vegetable, salt, wine, soya sauce and sugar, then add the crab fat. Fry for 3 minutes. Serve while it is hot.

碧玉珊瑚

材料：

半磅蟹黄，煮熟

三又二分之一磅芥蓝菜，选嫩茎部分，或者以其他的绿色蔬菜代之

一片生姜

三分之一杯猪油

一小匙细盐

三小匙米酒

四小匙酱油

一小匙白糖

做法：

首先将炒锅加油烧热。把姜略炒一下，去姜。再炒蔬菜，加盐、酒、酱油和白糖，然后加入蟹黄，炒三分钟。这道菜要很热时上桌供餐。

VEGETABLE DISHES
[Ts'ai Shu Lei(菜蔬类)]

Pages

204	Tientsin Cabbage with Chestnuts
206	Sour Tientsin Cabbage
208	Fried Mustard Greens
210	Braised Chinese Vegetables
212	Braised Fresh Mushrooms with Chinese Cabbage
214	Foreign Cabbage
216	Braised Bamboo Shoots
218	White Sauce Water Bamboo Shoots
220	Fried Green Bean Sprouts
222	Home-Made Pickles
224	Fried Spinach with Bamboo Shoots and Mushrooms
226	Braised Green Peas

菜蔬类

205　红烧栗子白菜

207　酸辣白菜

209　炒芥蓝菜

211　焖油菜

213　烩口蘑油菜

215　红烧洋白菜

217　油焖笋

219　白汁莴笋

221　炒豆芽菜

223　泡菜

225　炒菠菜冬菇笋

227　烩豌豆

TIENTSIN CABBAGE WITH CHESTNUTS
[Hung Shao Li Tzu Pai Ts'ai (红烧栗子白菜)]

Ingredients:

2	Tientsin cabbages (hearts only, cut into coarse pieces)
10	medium-sized mushrooms (soaked and cooked)
1	cup of chestnuts
1	teaspoonful of salt
1/4	cup of stock
6	teaspoonfuls of soya sauce
2	teaspoonfuls of sugar
1/2	cup of lard

Method:

Cut the chestnuts into halves with the shell, boil them for 10 minutes, and then remove the shells.

Place the lard in a frying pan, when very hot, put in the cabbage and mushrooms, and fry slowly for about 10 minutes. Then add the salt, sugar, soya sauce and stock, and cook all together for 5 minutes.

Finally add the chestnuts, and cook for a further 5 minutes.

红烧栗子白菜

材料：

二个白菜，只要菜心，切成大片

十只中型大小冬菇，用开水泡发，煮熟

一杯栗子

一小匙细盐

四分之一杯原汁肉汤

六小匙酱油

二小匙白糖

半杯猪油

做法：

首先将栗子切成对半，把它煮上十分钟，然后去壳。

把猪油置在炒锅中烧热，将菜和冬菇用慢火炒十分钟。然后加细盐、糖、酱油和原汁肉汤，混合之后煮上五分钟。

最后加那栗子肉，同时再煮五分钟。

SOUR TIENTSIN CABBAGE
[Suan La Pai Ts'ai (酸辣白菜)]

Ingredients:

1	lb. Tientsin cabbage (hearts only)
2	hot chilis (sliced)
1	teaspoonful of red pepper
1/2	teaspoonful of cornflour
1	teaspoonful of salt
2	teaspoonfuls of soya sauce
1	teaspoonful of sesamum oil
1/3	cup of vinegar (white)
1/2	cup of lard

Method:

Heat the lard in a pan. Fry the red pepper a little, then remove it.

Use the same fat to fry the chilis and cabbage for 3 minutes. Then add the rest of the ingredients, and cook for another 2 minutes. Serve warm.

酸辣白菜

材料：

一磅大白菜，只要菜心

二只辣椒，切碎

一小匙花椒

半小匙玉米粉

一小匙细盐

二小匙酱油

一小匙麻油

三分之一杯白醋

半杯猪油

做法：

将炒锅烧热倒上油。把花椒略炒数下，然后盛起。

用余下的油将辣椒和白菜炒三分钟。然后把那其他材料全部加入，再烧二分钟。要热的供餐。

FRIED MUSTARD GREENS
[Ch'ao Chieh Lan Ts'ai (炒芥蓝菜)]

Ingredients:

3½	lbs. Mustard Greens (hearts only)
1	big slice of ginger
1	teaspoonful of sugar
1/4	cup of lard
1	teaspoonful of salt
3	teaspoonfuls of wine
4	teaspoonfuls of soya sauce

Method:

Heat the lard in a frying pan. Fry the ginger a little, then remove the ginger. Add the vegetable, salt, wine, soya sauce and sugar. Stir the mixture continuously over a hot fire, and cook for 3 minutes. The vegetable should be crisp and green.

炒芥蓝菜

材料：

　　三又二分之一磅芥蓝菜，取菜心

　　一大片生姜

　　一小匙白糖

　　四分之一杯猪油

　　一小匙细盐

　　三小匙酒

　　四小匙酱油

做法：

　　将炒锅烧热加油。先把姜炒一下，然后盛起。再加菜、盐、酒、酱油和白糖。用很热的火候混合炒之三分钟。这样的做法是很脆的。

BRAISED CHINESE VEGETABLES
[Mên Yu Ts'ai (焖油菜)]

Ingredients:

- 2 lbs. Chinese cabbage hearts (Yu Ts'ai) (coarsely sliced)
- 2 tins of bamboo shoots (sliced thinly)
- 8 large mushrooms (soaked and cooked and cut up into halves)
- 1 slice of ginger
- 1½ teaspoonfuls of salt
- 1/2 cup of lard
- 1 cup of stock

Method:

Fry the ginger in hot lard, and discard the ginger. Then add the salt and vegetables. Fry just a little, mixing well together, and lastly add stock. Put all into a deep saucepan, and simmer for about 15 minutes.

焖油菜

材料：

二磅油菜心，切成大片

二罐竹笋(约一斤)，切成薄片

八只大冬菇，泡发，煮熟，切对开

一片生姜

一又二分之一小匙细盐

半杯猪油

一杯原汁肉汤

做法：

将油烧热，把姜炒一下，丢弃生姜。然后加盐和菜。略炒之，炒得很顺和之后，加入原汁肉汤。将全部材料转置在汤锅中，用慢火烧十五分钟，这道菜就算完成了。

BRAISED FRESH MUSHROOMS WITH CHINESE CABBAGE

[Hui K'ou Mo Yu Ts'ai (烩口磨油菜)]

Ingredients:

- 1 cabbage heart (boiled)
- 1 bamboo shoot (sliced)
- 1/2 lb. fresh mushrooms (cooked)
- 2 teaspoonfuls of cornflour
- 1/2 teaspoonful of salt
- 1/4 cup of lard
- 2 cups of water

Method:

Boil the mushrooms in about 2 cups of cold water for 5 minutes. Heat the lard in a frying pan, then add salt and the mushrooms together with the water they were cooked in. Then add the cabbage and bamboo shoot, and allow to cook for 10 minutes. Mix the cornflour with a little water and add it slowly to the mixture.

烩口蘑油菜

材料：

一个菜心，煮之

一只竹笋，切片

半磅鲜蘑菇，煮之

二小匙玉米粉

半小匙细盐

四分之一杯猪油

二杯水

做法：

用二杯水将蘑菇煮五分钟。再把油置在炒锅中烧热，加入细盐，蘑菇连水倒入烧之。然后加入菜和竹笋，煮十分钟。

用一点水将玉米粉调糊，徐徐地加进，熟时供食。

FOREIGN CABBAGE
[Hung Shao Yang Pai Ts'ai (红烧洋白菜)]

Ingredients:

- 2   cabbage hearts (broken up)
- 1   teaspoonful of salt
- 6   teaspoonfuls of soya sauce
- 1/2 cup of lard
- 2   teaspoonfuls of sugar

Method:

Heat the lard in a frying pan. Put in the cabbage and cook for 5 minutes. Then add the soya sauce, sugar and salt, and cook for a further 10 minutes, until the cabbage contains practically no water.

红烧洋白菜

材料：

二个洋白菜心，折碎

一小匙细盐

六小匙酱油

半杯猪油

二小匙白糖

做法：

首先将油烧热。把菜倒进，烧五分钟。然后加入酱油、白糖和盐，再烧十分钟，直到菜没有水分。好可供之。

BRAISED BAMBOO SHOOTS
[Yu Mên Sun (油焖笋)]

Ingredients:

- 10 bamboo shoots (cut up and boiled for 15 mins.)
- 1/2 teaspoonful of salt
- 1 heaped teaspoonful of sugar
- 5 teaspoonfuls of soya sauce
- 1 teaspoonful of wine
- 1/4 cup of lard
- 1/4 cup of stock

Method:

Heat the lard in a frying pan and fry the bamboo shoots (previously boiled) with the salt, sugar, soya sauce, wine and stock.

Turn out into a deep saucepan, bring to the boil, then simmer for 20 minutes.

Serve hot.

油焖笋

材料：

十只竹笋，切开，煮十五分钟

半小匙细盐

一小满匙白糖

五小匙酱油

一小匙酒

四分之一杯猪油

四分之一杯原汁肉汤

做法：

将油倒在炒锅中烧热，把已煮过的竹笋炒之，加入盐、糖、酱油、酒和原汁肉汤。

再将这材料转置在汤锅中，煮沸后用慢火煮廿分钟。

这道菜吃热的。

WHITE SAUCE WATER BAMBOO SHOOTS
[Pai Chih Wo Sun (白汁萵笋)]

Ingredients:

- 12 pieces of water bamboo shoots (using hearts only)
- 1/4 cup of lard
- 2 teaspoonfuls of flour
- 1 teaspoonful of salt
- 1 slice of ham (cooked and minced)
- 1/2 cup of meat stock

Method:

Boil the water bamboo shoots in cold water for 1/2 hour. Turn out onto a dish.

To make the sauce, put the lard in a hot frying pan. Add the flour and salt, and fry for a second. Then add the meat stock.

Pour this sauce over the water bamboo shoots in the dish and garnish with ham before serving.

白汁莴笋

材料：

十二根莴笋，只用菜心

四分之一杯猪油

二小匙面粉

一小匙细盐

一片火腿，煮熟，切碎

半杯原汁肉汤

做法：

将莴笋放入冷水煮半个小时，再把它盛在盘中。

做调味料时，先将油置在炒锅烧热，加面粉和细盐略炒数下，然后加入原汁肉汤。

这调味料做好之后倒在莴笋上面，再将已切碎的火腿装饰在上头供食。

FRIED GREEN BEAN SPROUTS
[Ch'ao Tou Ya Ts'ai (炒豆芽菜)]

Ingredients:

- 1   lb. bean sprouts
- 1   large stalk of celery (using heart only, shredded)
- 6   mushrooms (soaked and shredded)
- 1   teaspoonful of salt
- 4   teaspoonfuls of soya sauce
- 1/2 teaspoonful of cornflour
- 1/2 cup of lard
- 1/4 cup of stock

Method:

Heat the lard in a frying pan, add mushrooms and celery, and fry a little.

Then add the sprouts, salt, soya sauce and stock, and cook for 3 minutes.

Now add the cornflour mixed with a little cold water, stir into the vegetable and serve.

炒豆芽菜

材料：

　　一磅豆芽菜

　　一大根芹菜，只用心，切碎

　　六只冬菇，泡发，切丝

　　一小匙细盐

　　四小匙酱油

　　半小匙玉米粉

　　半杯猪油

　　四分之一杯原汁肉汤

做法：

　　将炒锅加油烧热，加入冬菇和芹菜，略炒几下。

　　然后加豆芽、盐、酱油和原汁肉汤，煮约三分钟。

　　现在用一点水将玉米粉调糊，徐徐地加入菜中，熟后即可供食。

HOME-MADE PICKLES
[P'ao Ts'ai (泡菜)]

Ingredients:

- 2 cucumbers (seeds removed)
- 1 large cauliflower (in coarse pieces)
- 1/2 foreign cabbage (in coarse pieces)
- 6 green peppers (seeds removed and cut into squares)
- 3 red chilis (seeds removed and cut into squares)
- 1 red carrot (skin removed and cut into squares)
- 2 cups of white vinegar
- 1/2 cup of sugar
- 1 teaspoonful of salt

Method:

Boil the cauliflower in a deep saucepan for 10 minutes, then add all the above ingredients except the vinegar, sugar and salt.

Now boil the vinegar and pour onto the vegetables, and allow them to soak for 20 minutes.

Turn the vegetables over and cover the saucepan with the lid. Pickle for 5 hours. Serve cold.

泡 菜

材料：

二根黄瓜，去籽

一个花菜，切大块

半个卷心菜，切大块

六个青椒，去籽切成方块

三个辣椒，去籽切成方块

一个红萝卜，去皮切成丁

二杯白醋

半杯白糖

一小匙细盐

做法：

首先将花菜用开水煮十分钟，然后将其他材料全部加入，但除了醋、糖和盐。

现在将醋等煮一下，然后淋在菜上，泡廿分钟。

将菜上下翻转，盖上锅盖。泡五个小时之后可供食。

FRIED SPINACH WITH BAMBOO SHOOTS AND MUSHROOMS
[Ch'ao Po Ts'ai Tung Ku Sun (炒菠菜冬菇笋)]

Ingredients:

- 1 lb. spinach
- 1/4 tin of bamboo shoots (cut into long thin slices)
- 10 mushrooms (soaked, cooked and cut into long thin slices)
- 1½ teaspoonfuls of salt
- 1/2 cup of lard
- 1/4 cup of stock

Method:

Heat the lard in a frying pan, add the salt and spinach, and fry for 2 minutes. Then put in the rest of the ingredients, and fry altogether for about 4 minutes.

炒菠菜冬菇笋

材料:

一磅菠菜

四分之一罐竹笋,切成长条

十只冬菇,泡发,煮熟,切成长条

一又二分之一小匙细盐

半杯猪油

四分之一杯原汁肉汤

做法:

将油倒在炒锅中烧热,加入盐和菠菜,炒二分钟。然后将其他材料全部加入,混合炒四分钟,可供之。

BRAISED GREEN PEAS
[Hui Wan Tou (烩豌豆)]

Ingredients:

- 1   tin of peas
- 1   bamboo shoot (cut into small squares)
- 8   small mushrooms (soaked, cooked and cut into small squares)
- 2   slices of Chinese ham (cut into small squares)
- 1   teaspoonful of salt
- 1½ teaspoonfuls of cornflour
- 1/3 cup of lard
- 1   cup of stock

Method:

Heat the lard in a hot frying pan, and add the bamboo shoot, mushrooms, peas, ham, salt and stock.

Mix the cornflour with a little cold water in a cup and add it slowly to thicken. Cook for about 5 minutes.

烩豌豆

材料：

一罐豌豆

一只竹笋，切成小丁

八只冬菇，泡发，煮熟，切成小丁

二片金华火腿，切成小丁

一小匙细盐

一又二分之一小匙玉米粉

三分之一杯猪油

一杯原汁肉汤

做法：

先将炒锅加油烧热，把竹笋、冬菇、豌豆、火腿、盐和原汁肉汤加入炒之。

然后用点水将玉米粉调糊徐徐加入。再烧五分钟可食之。

SOUPS
[T'ang Lei (汤类)]

Pages

230 Melon Soup

232 Duck Tongue Soup

234 Mushroom Soup

236 Button Mushroom Bean Curd Soup

238 Chicken and Cucumber Soup

240 Steamed Tientsin Cabbage Soup

242 Chicken, Mushroom and Bamboo Shoot Soup

244 Clear Duck Soup

汤 类

231　冬瓜盅
233　鸭舌汤
235　清炖花菇汤
237　口蘑豆腐汤
239　黄瓜余鸡汤
241　蒸天津白菜段
243　三鲜汤
245　清蒸鸭子汤

MELON SOUP

[Tung Kua Chung (冬瓜盅)]

Ingredients:

1/4	lb. pork fillet	
2	mushrooms (diced)	
1/4	lb. lotus seed	
1/4	lb. cooked ham	
1	stalk of Chinese onion	⎫
1	slice of ginger	⎬ fry together for 2 mins.
2	teaspoonfuls of salt	⎪
1	teaspoonful of wine	
1	tablespoonful of lard	
4	cups of meat sock	⎭
1	melon	

Method:

Cut the melon into halves like a bowl. Remove the seeds and spongy material from the melon. Cook it in cold water and bring to a boil.

Then remove it standing in a deep bowl. Put all the cooked ingredients into the melons, and steam for 3 hours.

冬瓜盅

材料：

四分之一磅猪里脊肉

二只冬菇，切丁

四分之一磅莲子

四分之一磅熟火腿

一根青葱

一片生姜

二小匙细盐

一小匙酒

一汤匙猪油

四杯原汁肉汤

将以上全部的材料炒二分钟。

一只冬瓜

做法：

首先将冬瓜切下像碗似的一半，把籽和心除去，放入凉水中煮开。

把冬瓜立至一个深碗，再加入以上全部其他材料，蒸上三个小时。

DUCK TONGUE SOUP
[Ya She T'ang (鸭舌汤)]

Ingredients:

- 24   duck tongues
- 1   small piece of ginger
- 1   small stalk of Chinese onion
- 1½   teaspoonfuls of salt
- 1   teaspoonful of Chinese wine
- 4   cups of meat stock

Method:

Steam the duck tongues with the ginger, onion, wine and 1/2 teaspoonful of salt for 1 hour.

Bring the meat stock to the boil in a deep saucepan. Add all the ingredients and another teaspoonful of salt. Serve in a deep bowl.

鸭舌汤

材料：

廿四条鸭舌

一小片生姜

一小根青葱

一又二分之一小匙细盐

一小匙酒

四杯原汁肉汤

做法：

先将那鸭舌同生姜、葱、酒和半小匙盐蒸一个小时。

再将原汁肉汤煮开,加入蒸熟的材料和另一小匙细盐。食时用大汤碗盛之。

MUSHROOM SOUP
[Ch'ing Tun Hua Ku T'ang (清炖花菇汤)]

Ingredients:

- 1/4 lb. mushrooms (soaked)
- 1/2 chicken
- 2½ teaspoonfuls of salt
- 6 cups of water
- 1 thin slice of ginger
- 1 stalk of Chinese onion
- 1 teaspoonful of wine

Method:

Chop the chicken into large-sized pieces with the bone, then boil it in the water for 3 hours over a slow fire together with the other ingredients.

When serving, use the soup and mushrooms only.

清炖花菇汤

材料：

四分之一磅冬菇，泡发

半只鸡

二又二分之一小匙细盐

六杯水

一薄片生姜

一根青葱

一小匙米酒

做法：

起先将鸡连骨切成大块，用慢火将鸡和其他所有的材料混合煮三个小时。

吃的时候只取汤和冬菇供食。

BUTTON MUSHROOM BEAN CURD SOUP
[K'ou Mo Tou Fu T'ang (口蘑豆腐汤)]

Ingredients:

- 10   dried mushrooms (K'ou Mo)
- 1/4 lb. bean curd (cut into squares)
- 1   teaspoonful of salt
- 4   cups of stock

Method:

Soak the mushrooms in boiling water for 10 minutes. Take them out and wash them in cold water 2 or 3 times. Then return the mushrooms to the original water in which they have been soaked.

Bring the stock to the boil in a frying pan.

Add the salt, bean curd and the mushrooms with their water, and bring to the boil a second time before serving.

口蘑豆腐汤

材料：

十只干蘑菇（口蘑）

四分之一磅豆腐，切成丁

一小匙细盐

四杯原汁肉汤

做法：

先把蘑菇用沸水泡开，大约要十分钟之久。再用凉水把蘑菇洗涤二至三道，然后将其放回原来泡发的水中。

把肉汤煮开，加盐、豆腐、蘑菇以及那些水，再煮开一回后供食。

CHICKEN AND CUCUMBER SOUP
[Huang Kua Ts'uan Chi T'ang (黄瓜汆鸡汤)]

Ingredients:

- 1 chicken (with bone)
- 4 cups of cold water
- 1 teaspoonful of salt
- 1/2 cucumber (sliced)
- 1/2 teaspoonful of cornflour

Method:

Boil the whole chicken (except the breast) in 4 cups of water for 2 hours to make the soup.

Slice the chicken breast, and mix it thoroughly with 1/2 teaspoonful of salt and the cornflour.

When the soup is ready, drain it off, using only the clear soup. Pour it into a frying pan, and boil. When boiling, add the cucumber, chicken breast and 1/2 teaspoonful of salt. Serve at once.

黄瓜汆鸡汤

材料：

一只鸡（连骨）

四杯水

一小匙细盐

半条黄瓜，切片

半小匙玉米粉

做法：

将整个鸡（除胸肉外）用四杯水煮上二小时。

把鸡胸肉切成片，与半小匙盐和玉米粉调和备用。

汤煮好之后，只取清汤，那鸡骨和其他的东西都不要。再把汤煮开加黄瓜、鸡胸肉和半小匙细盐煮之。趁热供食。

STEAMED TIENTSIN CABBAGE SOUP
[Chêng Pai Ts'ai Tuan (蒸天津白菜段)]

Ingredients:

- 2 large cabbage hearts (cut into big pieces)
- 1 slice of ham (finely chopped)
- 1 teaspoonful of salt
- 1 teaspoonful of cornflour
- 1½ cups of stock

Method:

Place the cabbage in a deep dish, and steam in a steam cooker for 15 minutes.

Make some sauce in a frying pan with the stock by adding salt and thickening with the cornflour. Pour this over the cabbage, and garnish with the finely chopped ham.

蒸天津白菜段

材料：

两大棵白菜心,切成大片

一片火腿,切碎

一小匙细盐

一小匙玉米粉

一又二分之一杯原汁肉汤

做法：

把菜装在大碗中,用蒸笼蒸上十五分钟。

将那些原汁肉汤加盐,以玉米粉增稠,在炒锅中做成调味汁,淋在菜的上头,再将那已切碎的火腿装饰在上面,供食。

CHICKEN, MUSHROOM AND BAMBOO SHOOT SOUP
[San Hsien T'ang (三鲜汤)]

Ingredients:

- 1 chicken (boiled in 6 cups of water for 1 ½ hours to make soup)
- 2 chicken breasts (sliced)
- 8 fresh mushrooms
- 1 large bamboo shoot (sliced)
- 1 teaspoonful of salt
- 1 teaspoonful of wine

Method:

Through a fine strainer strain the chicken soup into a hot frying pan. Bring to the boil, then add the chicken breast, fresh mushrooms, bamboo shoot, salt and wine.

Bring to the boil again, and serve in a deep bowl.

三鲜汤

材料:

一只鸡,用六杯水将它煮上一个半小时,做汤之用

二片鸡胸肉,切丝

八只鲜蘑菇

一大只冬笋,切片

一小匙细盐

一小匙酒

做法:

先用细滤器把鸡汤滤出放入热锅中,取那鸡的原汁汤,烧沸之后加入鸡胸肉丝、鲜蘑菇、冬笋、细盐和酒。

再把它煮开,用大汤碗装置供食。

CLEAR DUCK SOUP
[Ch'ing Chêng Ya Tzu T'ang(清蒸鸭子汤)]

Ingredients:

- 1   duck (about 3 lbs.)
- 1/2 bamboo shoot
- 4   big mushrooms (Tung Ku) (soaked)
- 1   piece of Chinese onion
- 1   piece of ginger
- 2   slices of cooked ham
- 1   heaped teaspoonful of salt
- 4   cups of water

Method:

Put the whole duck and all the other ingredients in a big bowl. Steam in a double boiler for 2 hours over a moderate fire. Serve in a soup bowl.

清蒸鸭子汤

材料：

一只鸭子（约三磅）

半只冬笋

四只大冬菇，泡发

一根青葱

一片生姜

二片熟火腿

一小满匙细盐

四杯水

做法：

将整个鸭子和全部其他材料置在大碗中。用蒸笼把它蒸上二个小时，用中火候为佳。供食时用大汤碗盛之。

DESSERTS
[T'ien Tien Hsin(甜点心)]

Pages

248—249	"Eight Precious" Rice Pudding
252	Almond Tea
254	Rose Petal Syrup Cake
256	Almond Curd
258	Walnut Tea

甜点心

250—251　八宝饭
253　　　杏仁茶
255　　　玫瑰锅炸
257　　　杏仁豆腐
259　　　核桃仁茶

"EIGHT PRECIOUS" RICE PUDDING
[Pa Pao Fan (八宝饭)]

Main Ingredients:

- 2 cups of glutinous rice (糯米)
- 1/2 lb. Chinese dates (cooked for 30 mins. and mashed)
- 2 ozs. dried red plums (红梅)
- 2 ozs. dried green plums (青梅)
- 2 ozs. candied lotus seeds (莲子)
- 2 ozs. honey dates (蜜枣)
- 2 ozs. chestnuts (cooked and shelled)
- 4 cups of cold water
- 1 cup of sugar
- 1 teaspoonful of lard

Sauce Ingredients:

- 1/2 cup of sugar
- 1 teaspoonful of cornflour
- 1 cup of cold water

Method:

Wash the rice 3 or 4 times in cold water. Put the washed rice and 4 cups of cold water in a deep saucepan, then boil it for about 5 minutes, and allow to simmer for 20 minutes. Now add the sugar and stir well.

Now line a 7-inch deep bowl with a piece of white paper cut into a round shape and smeared with lard on both sides. Place the red plums in the centre of the bowl, then around this, in order, arrange the green plums, the lotus seeds, the dates, and, on the outermost edge, the chestnuts. Now divide the rice into two equal portions. Place one portion over the dried fruit in the bowl. Cover this with a layer composed of the mashed dates, and, finally, place the other portion of rice on top of the dates. Place the bowl and its contents in a steamer, and steam for 20 minutes.

Boil the water and sugar in a saucepan, then mix the cornflour with a little cold water and add it slowly to the syrup.

Turn the rice pudding out onto a dish, remove the paper and pour the sauce over the pudding before serving.

八宝饭

主料：

二杯糯米

半磅红枣，煮三十分钟，搅成泥

二盎司[1]红梅干

二盎司青梅干

二盎司糖渍莲子

二盎司蜜枣

二盎司栗子，煮熟去壳

四杯水

一杯白糖

一小匙猪油

调味汁的材料：

半杯白糖

一小匙玉米粉

一杯水

1　1盎司等于28.35克。——编者注。

做法：

首先将米洗三或四次。洗好之后放四杯水，用深底的锅子来煮，先用大火候沸煮五分钟，然后用慢火候再煮廿分钟。加入白糖搅拌均匀。

现在取一个七英寸的汤碗，将一张白纸剪成和碗同样大小，两面搽上油，把纸粘在碗上。再把红梅排在碗底正中间，然后把青梅、莲子、蜜枣排在周围，把栗子排在最旁边。现在把饭分成二份。先用半数的饭加在已排好的百果上面，再将枣泥加入在当中，然后把另一半的饭加进，和碗边口平。将它转置在蒸笼中蒸之，约廿分钟之久。

把水和白糖在锅中煮之，然后用一点水将玉米粉调之，徐徐地加入，至沸后即可作为调味汁。

吃的时候把八宝饭翻在盘中，除去白纸，淋上调味汁。

ALMOND TEA
[Hsing Jên Ch'a (杏仁茶)]

Ingredients:

- 3/4 cup of uncooked rice (soaked in cold water for 15 mins.)
- 1/2 cup of sweet almonds (blanched in boiling water for 15 mins.)
- 1/4 cup of bitter almonds (soaked in boiling water for 15 mins.)
- 6 cups of cold water
- 2 cups of sugar

Method:

Put the rice and almonds in a mortar, and pound thoroughly adding a cup of water slowly.

Strain through a muslin bag, squeezing it well.

Add 5 cups of water, then 2 cups of sugar.

Bring to the boil, then boil gently for 7 minutes, stirring all the time.

Serve hot or cold.

杏仁茶

材料：

四分之三杯生白米，洗涤之后用凉水泡十五分钟

半杯甜杏仁，破开，用开水泡十五分钟

四分之一杯苦杏仁，用开水泡十五分钟

六杯凉水

二杯白糖

做法：

将米和杏仁倒入研钵，加入一杯水将它磨碎。

全部好了之后，用一个小棉布袋装上米和杏仁，把水压干，去渣。再加入五杯水和二杯白糖，将它煮沸后，接着用慢火煮七分钟，煮的时候不断地搅动。

这道点心吃冷的或热的均可。

ROSE PETAL SYRUP CAKE
[Mei Kuei Kuo Cha (玫瑰锅炸)]

Main Ingredients:

- 3 egg yolks
- 1 cup of cold water
- 1 tablespoonful of rose sugar (玫瑰糖)
- 1/2 cup of flour
- 2 teaspoonfuls of sugar

Syrup Ingredients:

- 4 teaspoonfuls of sugar
- 1/2 teaspoonful of cornflour
- 1 teaspoonful of rose petal syrup (玫瑰糖稀)
- 1 cup of cold water

Method:

Mix well together the egg, the flour, 2 teaspoonfuls of sugar and the water. Boil until it sets, then turn out and allow to cool.

Cut it into strips, 2 inches long and 1/2 inch wide. Sprinkle them with a little cornflour. Fry these in deep fat until light brown.

Sprinkle the rose sugar over them, and serve with the following syrup:

Bring the water to the boil in a small deep saucepan. Add the sugar and rose petal syrup. Mix the cornflour with a little water, and add it slowly to thicken the syrup.

玫瑰锅炸

主料：

三只鸡蛋的黄　　　　半杯面粉

一杯凉水　　　　　　二小匙糖

一大匙玫瑰糖

糖酱的材料：

四小匙白糖　　　　　一小匙玫瑰糖稀

半小匙玉米粉　　　　一杯凉水

做法：

首先将鸡蛋、面粉、二小匙白糖和凉水调得很顺和。煮之至定型，熟后将它倒出冷却。

再把它切成二英寸长、半英寸阔的长条形。用一点玉米粉撒附着，将油烧热把它炸成黄金色。

盛出，撒上玫瑰糖，吃的时候和如下制作之糖酱同食：

先用一个小锅子把水烧开，加入白糖和玫瑰糖稀。用一点水将玉米粉调和，再徐徐地加进沸水中煮之。约一分钟即盛出。

ALMOND CURD
[Hsing Jên Tou Fu (杏仁豆腐)]

Ingredients:

- 1/2 cup of sweet almonds (blanched in boiling water)
- 1/2 cup of uncooked rice (soaked in cold water)
- 1/4 cup of bitter almonds (soaked in boiling water)
- 2 strips of gelatine
- 1¾ cups of sugar
- 9 cups of water
- 6 cherries (cut into halves)

Method:

Put the almonds and the rice in a grinder, while grinding add 1 cup of water. Pour the liquid into a muslin bag, and squeeze out the watery portion which should be about one cup. To this add 5 cups of water, then add the gelatine and 1 cup of sugar. Boil until the gelatine has melted. Pour into a flat dish and allow to set. Cut into fairly large diamond shaped pieces.

Boil 3 cups of water, add remaining sugar, and boil for a few minutes. When cold, add to the sliced almond curd and decorate with the cherries.

杏仁豆腐

材料：

半杯甜杏仁，破开，泡在开水里

半杯生米，用凉水泡之

四分之一杯苦杏仁，泡在开水中

二包胶粉

一又四分之三杯白糖

九杯凉水

六只樱桃，切成对开

做法：

起先将杏仁和米混合置于研磨器中，磨的时候加一杯水。磨的出口置上一个布袋，挤压滤得液体部分，约为一杯水的量，然后加五杯水，再加胶粉和一杯白糖。将以上的材料煮之直到胶粉溶化。再倒在盘子中至它冻起。切成菱形。

将三杯水和剩下的白糖煮几分钟。煮好之后让糖水凉之，吃的时候淋在杏仁豆腐上面，再将樱桃装饰在最上面。

WALNUT TEA
[Hê T'ao Jên Ch'a (核桃仁茶)]

Ingredients:

- 2 cups of walnuts (shelled, blanched in boiling water for 15 mins. and skin removed)
- 1/2 cup of uncooked rice (soaked in cold water for 15 mins.)
- 2 cups of sugar
- 6 cups of water

Method:

Grind the walnuts and rice together thoroughly in a mortar, adding water, a little at a time, while grinding. Then add what remains of 3 cups of water to the walnut and rice paste. Squeeze through a muslin bag.

To the liquid thus obtained, add 3 more cups of water and the sugar, and cook in a deep saucepan for 7 minutes, stirring all the time. Serve in cups.

核桃仁茶

材料：

二杯核桃，剥壳，用开水泡约十五分钟，去皮

半杯生米，用凉水泡十五分钟

二杯白糖

六杯水

做法：

把核桃肉和米混合之，置于研钵中磨之，同时加水，一次加少量。然后把三杯水中用剩下的水加在核桃和米糊中。置在布袋里压干。

再把这液体加入三杯水和白糖，用汤锅将它煮上七分钟，煮的时候不断地搅动。食时用茶杯供之。

PASTRIES
[Mien Shih (面食)]

Pages

262	How to Make Noodles
264—265	Fried Noodles
268	Noodles in Chicken Soup
270—271	Steamed Dumplings
274—275	Fried Dumplings
278—279	Steamed Shao Mai
282—283	Spring Rolls
286—287	Fried Ravioli
290—291	Pancake Rolls

面 食

263	擀面
266—267	鸡丝炒面
269	鸡丝汤面
272—273	烫面饺子
276—277	锅贴饺子
280—281	蒸烧卖
284—285	炸春卷
288—289	炸馄饨
292—293	荷叶饼

HOW TO MAKE NOODLES
[Kan Mien (擀面)]

Ingredients:

    4   eggs (medium size)
    2½ cups of flour

Method:

Beat eggs slightly, mix well with the flour and knead into a soft dough. Cover with a damp cloth, and allow it to stand for 10 minutes.

Knead again for 5 minutes, sprinkle with cornflour, then roll out till very thin, about 20 inches wide. Fold into a number of pleats, then cut across the pleats finely as you would cut rashers of bacon. By lifting the ends in the uppermost pleat, you get the long strands of noodles. (see fig)

To cook, drop the noodles into a saucepan of boiling water and boil for 2 minutes. Turn them out into a colander, and run cold water through it 3 times. Squeeze out the water and drain.

These noodles can now be used for Chao Mien or Tang Mien.

擀　面

材料：

　　四只鸡蛋（中等大小）

　　二又二分之一杯面粉

做法：

　　首先将蛋打碎，和面粉揉得很顺和。再用湿的布包上，静置约十分钟之久。

　　再揉五分钟，然后筛点粉在上头，把它擀成薄片，这个面团约廿英寸那么宽就够薄了。再把它折起来，那么现在可以像切咸肉薄片那样切成面条。

　　煮的时候先将水烧开再下面条，大概要二分钟。用漏勺盛起，过三遍凉水，然后排干水。

　　这面条随您做炒面或汤面。

FRIED NOODLES
[Chi Ssu Ch'ao Mien (鸡丝炒面)]

Ingredients:

cooked noodles

1   lb. bamboo shoot (shredded)

1/4   lb. chicken breast (shredded)

3   stalks of Chinese celery (hearts only, shredded)

6   large mushrooms (soaked and shredded)

1/2   large onion (shredded)

1   cup of lard

1½   teaspoonfuls of salt

4   teaspoonfuls of soya sauce

1   egg (made into a thin omelet shredded)

1   cup of stock

1   slice of cooked ham (shredded)

Method:

Heat 1/2 cup of lard in a frying pan, add the noodles, sprinkle 1/2 teaspoonful of salt on them and fry one side until a golden brown. Turn over and repeat the process. Turn noodles out onto a dish.

Add 1/3 cup of lard, and fry onion to a light brown. Put in mushrooms and all the other ingredients (except egg omelet and ham), together with 1/2 teaspoonful of salt, 4 teaspoonfuls of soya sauce and 1 cup of stock, and fry for 3 to 4 minutes.

Divide the contents of the pan into two equal portions: leave one half in the pan, pour in the noodles, and mix together, then replace the other half on this mixture. Garnish with shredded ham and omelet.

Serve on a large dish while very hot.

Sufficient for 8 persons.

鸡丝炒面

材料：

煮过的面条

一磅竹笋，切丝

四分之一磅鸡胸肉，切丝

三根芹菜，只取菜心，切丝

六只大冬菇，泡发，切丝

半只大洋葱，切丝

一杯猪油

一又二分之一小匙细盐

四小匙酱油

一杯原汁肉汤

一只蛋，打碎煎成薄片，切丝

一片熟火腿，切丝

做法：

把半杯猪油倒在锅中烧热，将面加入，再撒上半小匙细盐，让它一面煎成金黄色。再将它反个面如上煎之。好了之后盛在盘子上。

再加入三分之一杯猪油在炒锅中，先将洋葱炒成金黄色。加入冬菇和其他全部的菜料（除蛋丝、火腿丝），同时加半匙细盐、四小匙酱油、一杯原汁肉汤，再烧三至四分钟。

把这作料分成两份：一半留在炒锅中，将那炒好的面条加进调之，另一半的作料加在上头，以火腿丝及蛋丝装饰之。

趁热盛在大盘子上供食。这可供八人份。

NOODLES IN CHICKEN SOUP
[Chi Ssu T'ang Mien (鸡丝汤面)]

Ingredients:

noodles

1   chicken (for making soup)

2   pieces of chicken fillet (shredded)

2   bamboo shoots (shredded)

4   mushrooms (soaked and shredded)

1   bowl of cooked green vegetables (any kind)

1   teaspoonful of soya sauce

1½   teaspoonfuls of salt

1/8 cup of lard

Method:

Drop the noodles into boiling water, and bring to a boil. Remove them, and put them into the chicken soup, which should be boiling hot, adding 1 teaspoonful of salt. Turn out into a big bowl, or serve the noodles in individual small bowls.

Heat the lard in a frying pan. Fry the chicken fillet a little, then add the bamboo shoot, mushrooms, vegetables, soya sauce and 1/2 teaspoonful of salt, and fry together for 3 minutes. Turn out on top of the noodles in the bowl, or on top of each small bowl.

鸡丝汤面

材料:

面条

一只鸡,做成汤

二片鸡胸肉,切丝

二只冬笋,切丝

四只冬菇,泡发,切丝

一碗煮过的青菜(任何一种)

一小匙酱油

一又二分之一小匙细盐

八分之一杯猪油

做法:

先将面条煮熟。盛出面条放在热鸡汤里,加入一小匙细盐,倒在大汤碗中或者分开盛于饭碗中。

用一点油将鸡丝略炒一下,再加冬笋、冬菇、青菜、酱油和半小匙细盐,一同炒之约三分钟。盛在大碗或小碗的面条上供餐。

STEAMED DUMPLINGS
[T'ang Mien Chiao Tzu (烫面饺子)]

Ingredients for Wrapping:

- 2 cups of flour
- 1 cup of boiling water

Ingredients for Filling:

- 1/4 cup of lard
- 12 mushrooms (diced)
- 10 water chestnuts (diced)
- 1 bamboo shoot (diced)
- 1/2 lb. pork (minced)
- 1 teaspoonful of salt
- 5 teaspoonfuls of soya sauce
- 2 teaspoonfuls of sugar
- 1 tiny piece of Chinese onion (finely chopped)
- 1 tiny slice of ginger (finely chopped)
- 1 teaspoonful of wine
- 1 teaspoonful of cornflour
- 1 teaspoonful of sesamum oil

Method:

First make filling. Heat the lard in frying pan. Fry the pork together with the mushrooms, water chestnuts, bamboo shoot, onion, ginger, wine and sesamum oil. Then add the soya sauce, sugar and salt. Mix the cornflour with the cold water in a cup and pour over the pork, stir well, and turn out into a bowl.

To make wrapping mix the flour with boiling water until it forms a soft dough. Knead well, sprinkle with dry flour, and roll into a long sausage. Pinch off small pieces of uniform size, and with a small rolling pin roll each piece out into a circular shape about 3 inches in diameter, taking care to leave the centre thicker than the edge.

Place on the middle of each circular piece of dough about 1 teaspoonful of the filling mixture. Fold over to make a semicircle, then press the opposite edges together with the fingers, and you have your Chiao Tzu.

When all the Chiao Tzu are made, place them on a piece of damp cloth in a steam cage, and steam for about 10 minutes.

烫面饺子

面皮的材料：

　　二杯面粉

　　一杯开水

内馅的材料：

　　四分之一杯猪油

　　十二只冬菇，切碎

　　十只荸荠，切碎

　　一只竹笋，切碎

　　半磅猪肉，剁碎

　　一小匙细盐

　　五小匙酱油

　　二小匙白糖

　　一小根青葱，切碎

　　一小片生姜，切碎

　　一小匙酒

　　一小匙玉米粉

　　一小匙麻油

做法：

  首先制馅。将油置在炒锅里烧热。把那猪肉和冬菇、荸荠、竹笋、葱、姜、酒和麻油混合炒之。然后加酱油、白糖和盐。用一点水和玉米粉调糊，徐徐地倒在作料里搅之，熟后盛在碗里。

  第二步做面皮。面粉和开水调得很顺和，再将它揉之，筛点面粉把它揉成长条。然后切成小块，擀成直径三英寸的圆形薄片，中间较厚点。

  在圆形薄片中间放一小匙馅料，将之折叠成半圆，然后用手指将两边捏紧，就将它做成了饺子。

  全部做好了之后，在蒸笼底加一块湿白布，然后将饺子排列在上面蒸上十分钟。

FRIED DUMPLINGS
[Kuo T'ieh Chiao Tzu (锅贴饺子)]

Ingredients for Wrapping:

- 2 cups of flour
- 30 teaspoonfuls of cold water

Ingredients for Fillings:

- 3/4 lb. pork (minced)
- a little Chinese onion (chopped finely)
- a little ginger (chopped finely)
- 1/2 cup of lard
- 1 teaspoonful of salt
- 3 teaspoonfuls of soya sauce
- 2 teaspoonfuls of sesamum oil

Method:

To the pork add the onion, ginger, salt, soya sauce and sesamum oil, and mix thoroughly together with a pair of chopsticks.

Mix 2 cups of flour with 1 cup of cold water and knead into a soft dough. Cover with a damp cloth, and allow to stand for 15 minutes before using.

Roll out the dough on a board into a long sausage, then pinch off pieces about the size of a walnut.

Sprinkle flour on each one, and with a small rolling pin roll it into a thin pancake about 3 inches in diameter. Into each of these pieces put 1 teaspoonful of the filling mixture. Fold the pancake over, and pinch the edges together with the fingers.

To make the fried dumplings put the lard in a frying pan, arrange the dumplings in rows in the pan, and fry the bottom to a light brown. Now pour 1/2 cup of cold water over them. Cover with a lid, and continue cooking until the water has dried up.

锅贴饺子

面皮的材料：

  二杯面粉

  卅小匙凉水

内馅的材料：

  四分之三磅猪肉，剁碎

  一点青葱，切碎

  一点生姜，切碎

  半杯猪油

  一小匙细盐

  三小匙酱油

  二小匙麻油

做法：

　　用筷子把肉、葱、姜、盐、酱油和麻油等材料混合调之。

　　将二杯面粉和一杯凉水混合调顺揉之。再将它用湿的白布包之，让它放置十五分钟。然后将面团揉成长条，再切成核桃那么大。在每一个上面撒上面粉，用一个小擀面杖擀成直径三英寸圆形。每一个填一小匙馅料在当中。将之折叠成半圆，然后用手指将两边捏紧，就把它做成了饺子。

　　煎锅里先加一点油，把已做好的饺子排列在煎锅里，将它们的底部煎成金黄色。最后加进半杯凉水，盖好锅盖，烧到无水分时这锅贴饺子即告完成，即可供食。

STEAMED SHAO MAI
[Chêng Shao Mai (蒸烧卖)]

Ingredients for Wrapping:

- 2 cups of flour
- 1 cup of boiling water

Ingredients for Filling:

1	lb. pork (minced)	
1	small slice of ginger (finely chopped)	
1	small stalk of Chinese onion (finely chopped)	mix together thoroughly, and leave in a bowl for use as filling
6	mushrooms (soaked and finely chopped)	
1	teaspoonful of sesamum oil	
1/2	teaspoonful of salt	
3	teaspoonfuls of soya sauce	
1/8	teaspoonful of sugar	
4½	tablespoonfuls of cold water	

Method:

To make wrapping mix 2 cups of flour with 1 cup of boiling water until it forms a soft dough. Knead well, and roll into a long sausage, then pinch off pieces of uniform size. With a small rolling pin roll each of these into flat cakes about 3 inches in diameter, leaving the centre thicker than the edge.

Fray the edge of each by first sprinkling flour over it, and scraping with the lip of a bowl.

Put about 2 teaspoonfuls of the filling mixture on the centre of each wrapper. Manipulate the wrapping dough so that it closes around the filling material, leaving the edge free, and you have your Shao Mai. Repeat the process until the filling is used up.

Steam the Shao Mai in a steam cage or double boiler for 15 minutes.

蒸烧卖

面皮的材料：

　　二杯面粉

　　一杯开水

内馅的材料：

　　一磅猪肉，剁碎

　　一小片生姜，切碎

　　一小根青葱，切碎

　　六只冬菇，泡发，切碎

　　一小匙麻油

　　半小匙细盐

　　三小匙酱油

　　八分之一小匙白糖

　　四又二分之一汤匙凉水

　　把以上的材料在大碗中搅得很顺。

做法：

先做面皮。用二杯面粉和一杯开水混合之，揉得很顺，再做成长条，然后切成和核桃那么大小。再用一个小擀面杖把它擀成直径三英寸的圆形，中央要厚点。

把每一个圆形面皮边缘撒上面粉，然后用碗边刮出荷叶裙边的样子。

接着包烧卖。约二小匙的馅料置在面皮的当中。将它包成圆形，将边缘自由散开，这就做好了烧卖。重复这个过程直到馅料用完为止。

排列在蒸笼中，蒸上十五分钟之久可也。

SPRING ROLLS
[Cha Ch'un Chüan (炸春卷)]

Ingredients for Wrapping:

- 2 cups of flour
- 1½ cups of water

Ingredients for Filling:

- 1/4 lb. chicken breast (shredded)
- 4 mushrooms (shredded)
- 1/2 onion (shredded)
- 1 bamboo shoot (shredded)
- 2 cups of bean sprouts
- 2 cups of spinach in 1 inch lengths
- 1 teaspoonful of cornflour
- 2 teaspoonfuls of soya sauce
- 1 teaspoonful of Chinese wine
- 1/4 teaspoonful of salt
- 1/2 cup of lard
- 1/4 cup of stock

Method:

Add the salt and 1/2 teaspoonful of cornflour to the chicken and mix well. Put 1/4 cup of lard in a pan. When hot, fry the onion just a little. Add the mushrooms, bamboo shoot, bean sprouts, spinach and 1 teaspoonful of soya sauce, and fry together for 3 minutes. Turn out onto a plate. Place a little more lard in the pan, then fry the chicken meat with 1 teaspoonful of wine and 1 teaspoonful of soya sauce. Mix well, then add the other ingredients with 1/2 teaspoonful of cornflour stirred with stock or water.

The above is to be used as filling.

To make wrapping for spring rolls take 2 cups of flour, add cold water gradually and mix until the resulting dough is of the consistency of marshmallow, then let it stand for 10 minutes before using. Take the mass in your hand, and by pressing it against a warm greasy frying pan with a flat surface, a thin sheet is left on it which resembles a sheet of rice-paper, about the size of the palm of the hand. Repeat the process until all the dough is used up.

Into each piece put 2 teaspoonfuls of the above filling. Roll up lengthwise and seal the ends with cold water. Fry in deep fat till a golden brown.

炸春卷

面皮的材料：

　　二杯面粉

　　一又二分之一杯水

内馅的材料：

　　四分之一磅鸡胸肉，切丝

　　四只冬菇，切丝

　　半只洋葱，切丝

　　一只冬笋，切丝

　　二杯豆芽

　　二杯菠菜，切成一英寸长

　　一小匙玉米粉

　　二小匙酱油

　　一小匙黄酒

　　四分之一小匙细盐

　　半杯猪油

　　四分之一杯原汁肉汤

做法：

用盐和半小匙玉米粉混合在鸡肉内调之。倒四分之一杯猪油在锅子里，锅子热时将洋葱略炒几下。再加冬菇、冬笋、豆芽、菠菜和一小匙酱油，一同炒三分钟。然后将作料盛在盘中。再加点油在炒锅里，然后把鸡肉炒之，同时加一小匙酒和一小匙酱油。最后加入刚才炒好的作料，半小匙玉米粉用点原汁肉汤或水调顺徐徐加入。以上是春卷的馅料。

接着做春卷皮。用二杯面粉，加入凉水搅调很顺和，直到这种面像薄浆糊似的。让它停留十分钟才能用。用手捉一把面团贴糊在温热略带油的平底铁煎锅上，转成十英寸圆形薄片。等那春卷皮的周围立起时即可用手将它撕起。重复这一操作直到面团用完。

做春卷时每一张的当中加二小匙馅料。包的时候将长边卷起，用一点水将两端严封。

用较多的油把它炸成金黄色即可供食。

FRIED RAVIOLI
[Cha Hun T'un (炸馄饨)]

Ingredients for Wrapping:

    1   cup of flour

    2   eggs (small size)

Ingredients for Filling:

    1/2 lb. chicken (minced)
    1   small onion (finely chopped)
    1   thin slice of ginger (finely chopped)
    1   teaspoonful of salt
    1   teaspoonful of soya sauce
    1   teaspoonful of sesamum oil
    1   teaspoonful of Chinese wine

} mix thoroughly in a bowl

Method:

Mix the flour and eggs well and knead into a dough. Sprinkle cornflour on rolling board, roll out till very thin about 20 inches wide, and then cut into strips about 2½ inches in width. Place the strips one on top of the other, and cut them up into "isosceles trapezoids".

Onto each of these pieces of dough, place half teaspoonful of the filling towards the shorter side (S). Turn this side over the filling and roll halfway up towards the longer side (L). Double this over on itself, and pinch the ends of the rolled-up portion together, leaving the longer side free. The filling should be suffcient for 50 to 60 ravioli.

Fry in deep fat until a golden brown.

炸馄饨

面皮的材料：

　　一杯面粉

　　二只较小的鸡蛋

内馅的材料：

　　半磅鸡肉，切碎

　　一只小洋葱，切碎

　　一薄片生姜，切碎

　　一小匙细盐

　　一小匙酱油

　　一小匙麻油

　　一小匙酒

　　将以上的材料置在大碗中搅得很顺和。

做法：

将面粉和鸡蛋搅得很顺和后揉成面团。再把面团放在面板上揉得很调和，筛上一些粉将之擀成廿英寸那么大的薄片然后切成条，约二又二分之一英寸宽。把条状面片叠成一撂，切成"等腰梯形"。

将每一张馄饨皮朝短边处盛半小匙内馅料。把短边盖住馅料，折到朝长边一半处。再把包住部分向上折一次，将卷起部分的末端捏在一起，留下一点长边。这馅料可以做成五十至六十个馄饨。

用较多点油将它炸成金黄色供食。

PANCAKE ROLLS
[Ho Yeh Ping (荷叶饼)]

Ingredients for the Pancakes:

- 2 cups of flour
- 1/4 cup of lard
- 1 cup of boiling water

Ingredients for Filling:

- 1/2 lb. chicken (shredded)
- 1½ lbs. bean sprouts
- 1 stalk of celery (shredded)
- 3 mushrooms (soaked and shredded)
- 1/2 foreign onion (shredded)
- 1 teaspoonful of salt
- 5 teaspoonfuls of soya sauce
- 1/2 teaspoonful of sugar
- 1/3 cup of lard
- 1/2 teaspoonful of cornflour

Method:

Mix the flour with boiling water and knead into a soft dough. Shape the dough into a long sausage, then pinch off small pieces of uniform size. Roll each of these between the palms of your hands, and press it flat. Take two of these and brush a little lard between them, and roll them out 5 inches in diameter. Fry them on both sides in a dry flat pan, then tear them apart, and they are ready for serving.

Heat 1/4 cup of lard in a frying pan. Brown the onion and add the bean sprouts, celery, mushrooms and salt. Allow to cook for 2 minutes, and finally add the sugar and 4 teaspoonfuls of soya sauce. Turn out onto a dish.

Mix chicken with cornflour and salt thoroughly. Put the rest of the lard in the pan. Fry the chicken and add 1 teaspoonful of soya sauce. Replace the bean sprouts mixture in the pan. Fry for 2 minutes. Serve with the pancakes.

荷叶饼

饼皮的材料：
- 二杯面粉
- 四分之一杯猪油
- 一杯开水

内馅的材料：
- 半磅鸡肉，切碎
- 一又二分之一磅豆芽菜
- 一根芹菜，切碎
- 三只冬菇，泡发，切丝
- 半个洋葱，切丝
- 一小匙细盐
- 五小匙酱油
- 半小匙白糖
- 三分之一杯猪油
- 半小匙玉米粉

做法：

将面粉和开水混合搅成柔软的面团。将面团揉成长条，然后把它捏成大小均匀的小块。把每个小块放在手掌之间，然后按扁。拿两个，在中间刷一点猪油，把它们擀成直径五英寸的大小。用干的平底锅将它们两面煎好，然后撕开，即可供食。

四分之一杯猪油倒在炒锅中烧热。将洋葱炒成金黄色，再加入豆芽菜、芹菜、冬菇和细盐。让它烧二分钟，再加入糖和四小匙酱油。然后盛在碗里。

将鸡肉和玉米粉、盐调和。把那剩下的猪油置在炒锅里。把鸡肉炒之，再加一匙酱油。现在把已炒好的豆芽等再倒入炒上二分钟。这馅料和饼同时供餐。

Suggested Menus

(1)

Fried Rice

Fried Beef Fillet

Roast Crisp Duck

Red Sauce Rump of Pork

Braised Shrimps

Fried Mustard Greens

Steamed Tientsin Cabbage Soup

(2)

Fried Noodles

Roast Stuffed Chicken

Egg and Beef Omelets

Brown Sauce Mandarin Fish

Fried Spinach with Bamboo Shoots and Mushrooms

Velvet Chicken with Corn

(3)

Fried Dumplings

Minced Pigeon

Stuffed Green Peppers

Fried Fish Cakes

Stewed Rice Flour Pork

Tientsin Cabbage with Chestnuts

Melon Soup

Served with Rice

(4)

Spring Rolls

Sweet and Sour Pork

Fried Chicken with Pepper and Brown Sauce

Smoked Fish

Stewed Meatballs with Shantung Cabbage

Braised Green Peas

Duck Tongue Soup

Served with Rice

(5)

Fried Ravioli

Walnut Chicken

Fried Minced Fish

Brown Sauce Beef

Velvet Chicken (Imitation)

Braised Fresh Mushrooms with Chinese Cabbage

Chicken and Cucumber Soup

Served with Rice

(6)

Noodles in Chicken Soup

Red Sauce Duck

White Sauce Fish (Sour)

Fried Duck Liver

Roast Shaslick Pork

Stuffed Mushrooms

Home-Made Pickles

Served with Rice

(7)

Pancake Rolls

Fried Shrimp Balls

Fried Green Bean Sprouts

Pineapple and Ginger Duck

Stewed Pork with Bamboo Shoots

Mushroom Soup

Almond Tea

(8)

Roast Stuffed Chicken

Gold Coin Chicken

Fried Fish with Red Sauce

Stewed Shin of Beef

Steamed Dumplings

White Sauce Water Bamboo Shoots

Chicken, Mushroom and Bamboo Shoot Soup

建议菜单

(9)

Stewed Chestnut Chicken
Steamed Shao Mai
Crab Fat with Green Vegetables
Stewed Pork with Bamboo Shoots
Foreign Cabbage
Served with Rice
Walnut Tea

(10)

Steamed Dumplings
Chili Oil Chicken and Spinach
Shrimp Cakes
Crab Omelet
Roast Chicken (Boneless)
Fried Wild Duck
Braised Bamboo Shoots
Clear Duck Soup
Served with Rice
Rose Petal Syrup Cake

建议菜单

（1）

什锦炒饭

炒牛里脊

脆皮鸭子

扒肘子

烩虾仁

炒芥蓝菜

蒸天津白菜段

（2）

鸡丝炒面

烤酿鸡

煎鸡蛋饺

红烧鳜鱼

炒菠菜冬菇笋

鸡蓉玉米

（3）

锅贴饺子

鸽子松

牛肉酿辣椒

煎鱼饼

米粉肉

红烧栗子白菜

冬瓜盅

米饭

（4）

炸春卷

糖醋排骨

酱泡鸡丁

熏鱼

红烧狮子头

烩豌豆

鸭舌汤

米饭

（5）

炸馄饨

核桃鸡丁

炒鱼松

锅烧牛肉

芙蓉假鸡片

烩口蘑油菜

黄瓜余鸡汤

米饭

（6）

鸡丝汤面

红烧扒鸭

五柳鳜鱼

炒鸭肝

叉烧肉

酿冬菇

泡菜

米饭

（7）

荷叶饼

炸虾球

炒豆芽菜

菠萝姜鸭子

红焖五花猪肉

清炖花菇汤

杏仁茶

（8）

烤酿鸡

金钱鸡

红烧鱼

红烧牛肉

烫面饺子

白汁莴笋

三鲜汤

（9）

栗子焖鸡

蒸烧卖

碧玉珊瑚

红焖五花猪肉

红烧洋白菜

米饭

核桃仁茶

（10）

烫面饺子

辣油鸡丁

煎虾饼

芙蓉蟹肉

锅烧鸡

炒水鸭片

油焖笋

清蒸鸭子汤

米饭

玫瑰锅炸

MEANING （含义）	PICTOGRAPH （象形字）	CHARACTER （字符）	PRONUNCIATION （发音）
Saltish	鹹	鹹	Hsien
Sweet	甜	甜	T'ien
Sour	酸	酸	Suan
Bitter	苦	苦	K'u
Acrid	辣	辣	La
Fragrant	香	香	Hsiang

Chinese Characters for Different Tastes
不同口味的汉字

INDEX FOR CHINESE WORDS

熬 Ao	(Simmering)	12
炸 Cha	(Fry)	11
炒 Ch'ao	(Fry)	11
炒菜 Ch'ao Ts'ai	(Fried dishes)	18
蒸 Chêng	(Steam)	10
芡 Ch'ien	(Coating of flour)	14
请 Ch'ing	(Please)	41
酬 Ch'ou	(Return of compliment)	42
饭 Fan	(Rice)	20
香片 Hsiang P'ien	(Tea)	61, 84
花雕 Hua Tiao	(Wine)	64
烤炉 K'ao Lu	(Roasting oven)	8
炕 K'ang	(Big chair)	38
菇 Ku	(Mushroom)	73, 74

干杯 Kan Pei	(Dry cup)	41
焖 Mên	(Stew)	12
明炉 Ming Lu	(Open oven)	8
拼盆 P'ing P'ên	(Assorted dishes)	18
便饭 Pien Fan	(Ordinary meal)	20
煲 Po	(Boil)	13
烧烤 Shao K'ao	(Roast)	8
烧饭 Shao Fan	(Cook rice)	86
随便 Sui Pien	(No ceremony)	42

INDEX

Almond curd, 256

Almond tea, 252

Ball, shrimps, 194

Bamboo shoots, 76, 224

Bamboo and pork, 162

Bamboo, braised, 216

Bamboo, white sauce, 218

Bean sprouts, 220

Beef, brown sauce, 144

Beef, fillet, 135

Beef omelet, 138

Beef, stew, 146

Boiler, double, 12

Bowl, how to hold, 48

Braised bamboo shoot, 216

Braised green peas, 226

Braised shrimps, 188

Brown sauce beef, 144

Brown sauce mandarin fish, 167

Cabbage, Chinese, 212

Cabbage, foreign, 214

Cabbage, Shantung, 154

Cabbage, sour, 206

Cabbage, Tientsin, 204

Cake, fish, 184

Cake, rose petal, 254

Cake, shrimp, 192

Index 索 引

Chestnuts and cabbage, 204
Chicken, stewed chestnut, 100
Chicken, fried, 102
Chicken, gold coin, 94
Chicken, mushroom, 112
Chicken, roast, 108
Chicken, chili oil and spinach, 98
Chicken, stuffed, 106
Chicken, velvet, 104
Chicken walnut, 96
Chopper, 70
Chopsticks, 70
Chopsticks, how to hold, 48
Cooker, steam, 70
Clams, stuffed, 196
Crab omelet, 198
Crab fat with green vegetables, 200
Curd, almond, 256
Duck, roast crisp, 118
Duck, fried, 126
Duck, ginger, 116
Duck, pineapple, 116

Duck, red sauce, 120
Duck soup, 232, 244
Dumpling, fried, 274
Dumpling, steamed, 270
Egg and beef omelet, 138
Fish, 14
Fish, brown sauce, 167
Fish cake, 184
Fish, fried, 170, 178
Fish, minced, 178
Fish, sour, 170, 182
Fish, smoked, 174
Flavoring, 13
Flour, 14
Fried beef, 135
Fried chicken, 102
Fried bean sprouts, 220
Fried duck liver, 122
Fried duck, wild, 126
Fried fish with sour sauce, 170
Fried fish cake, 184
Fried shrimp ball, 194
Fried omelets, 138
Fried mustard greens, 208

Frying, 11

Frying pan, 70

Haw, red, 76

Ingredients, 14

Kitchen utensils, 70

Ladle, 70

Liver, red duck, 122

Meatball, 154

Melon soup, 230

Minced pigeon, 128

Mushrooms, 74, 224

Mushrooms, kinds of, 74

Mushrooms, to boil, 74

Mushrooms, stuffed, 112

Mushrooms and cabbage, 212

Noodles, how to make, 262

Noodles, fried, 264

Noodles, in chicken soup, 268

Omelet, crab, 198

Omelet, egg and beef, 138

Peas, braised, 226

Pigeon, minced, 128

Pickles, 222

Plain meals, 20

Pork, stewed rice flour, 158

Pork, red sauce, 152

Pork, roast shaslick, 160

Pork, stewed, 162

Pork, sweet and sour, 148

Pepper, green stuffed, 140

Presenting, 42

Ravioli, 286

Red haws, 76

Red sauce pork, 152

Red sauce duck, 120

Restaurant dinners, 18

Rice bowls, 48

Rice, how to cook, 86

Rice, fried, 88

Roll, spring, 282

Roll, pancake, 290

Rolling board, 70

Rolling pins, 70

Rose petal syrup cake, 254

Roast chicken, 106, 108

Roast crisp duck, 118

Roast shaslick pork, 160

Roasting, 8

Index 索 引

Saucepan, 70
Sautéing, 11
Seating, 38
Serving, 15
Shaohsing wine, 73
Shao Mai, steamed, 278
Shaslick, 160
Shrimps, 188, 192
Simmering, 12
Smoked fish, 174
Sour sauce fish, 170, 182
Soya sauce, 73
Spinach, 98, 224
Sprouts, bean, 220
Steaming, 10
Stewing, 12
Stewed beef, 146
Stewed chestnut chicken, 100
Stewed meatballs, 154
Stewed rice flour pork, 158
Stewed pork with bamboo shoot, 162
Strainer, 70
Stuffed clam, 196
Stuffed chicken, 106
Stuffed pepper, 140
Stuffed mushroom, 112
Soup, bean curd, 236
Soup, cabbage, 240
Soup, chicken, 238, 242
Soup, clear duck, 244
Soup, duck tongue, 232
Soup, melon, 230
Soup, mushroom, 234, 236
Table don'ts, 45
Table manners, 37
Table service, 57
Tea, 60
Tea, almond, 252
Tea, how to make, 84
Tea, walnut, 258
Utensils, 71
Walnut tea, 258
White sauce water bamboo shoot, 218
White sauce fish, 182
Wine, 64

索 引

熬，29
扒肘子，153
白菜，156，205
白汁莴笋，219
碧玉珊瑚，201
菠菜，99，225
菠萝鸭子，117
菜刀，72
餐仪之禁，54
餐桌礼仪，51
餐桌食具，59
叉烧肉，161
茶，62
炒，27
炒鱼，180
炒豆芽菜，221
炒鸡丁，103
炒芥蓝菜，209
炒面，266
炒牛里脊，137

炒勺，72
炒水鸭片，127
炒鸭肝，124
持碗，56
炊具，72
春卷，284
脆皮鸭子，119
冬菇，78—79，225
冬瓜盅，231
豆腐汤，237
豆芽菜，221
芙蓉假鸡片，105
芙蓉蟹肉，199
擀面，263
擀面板，72
擀面棍，72
鸽子松，130
烤鸡，107，109
锅烧牛肉，145
锅贴饺子，276

荷叶饼，292

核桃鸡丁，97

核桃仁茶，259

红焖五花猪肉，163

红烧鱼，168

红烧扒鸭，121

红烧鳜鱼，168

红烧牛肉，147

红烧狮子头，156

红焖五花猪肉，163

烩豌豆，227

烩虾仁，190

馄饨，288

鸡蛋饺，139

鸡丝汤面，269

鸡汤，239，243

家常便饭，35

煎，28

煎鸡蛋饺，139

煎鱼饼，186

姜鸭子，117

酱油，78

金钱鸡，95

酒，68

酒席，33

口蘑油菜，213

筷子，72

辣油鸡丁，99

栗子白菜，205

栗子焖鸡，101

漏勺，72

玫瑰锅炸，255

焖，28

焖猪肉，163

米粉肉，159

面粉，29

蘑菇种类，79

酿冬菇，114

酿海蛤蜊，197

酿鸡，107

酿辣椒，142

牛里脊，137

泡菜，223

菇汤，235，237

清蒸鸭子汤，245
砂锅，72
山楂饼，79
烧饭，87
烧烤，26
绍兴酒，78
什锦炒饭，90
狮子头，156
侍餐，30
手勺，72
松鼠鱼，172
酸辣白菜，207
糖醋排骨，150
烫面饺子，272
调味，29
碗，56
五柳鳜鱼，183
虾，190，193
虾饼，193
虾球，195
献，53
香片茶的泡法，85

杏仁茶，253
杏仁豆腐，257
选料，30
熏鱼，176
鸭舌汤，233
鸭汤，233，245
洋白菜，215
油焖笋，217
鱼，30
鱼饼，186
鱼松，180
炸松鼠鱼，172
炸虾球，195
蒸，27
蒸笼，72
蒸烧卖，280
蒸天津白菜段，241
执箸，55
竹笋，79，225
煮蘑菇汤，79
座位安排，52

AFTERWORD

I hope you have enjoyed this book and will take its wisdom to heart. This journey into the essence of Chinese culture should inspire Chinese readers to be proud of their heritage, while our foreign friends can now more fully appreciate the allure of the Chinese way of life.

Few if any cultures have a code of social etiquette and table manners that traces back 3,000 years. This ancestral code may have something to do with the fact that while other civilizations and kingdoms have come and gone, the Chinese people have persevered over the millennia. The persistence of our culture is certainly something that young Chinese can rightly take pride in, and the teachings that have guided us across all these centuries are something that in changing times all Chinese can find comfort in.

Chinese cuisine is at the heart of our culture and has served as our country's best ambassador, opening doors in every corner of the globe. I wish our wise readers to reflect on what the best of Chinese culture has to offer, not only to one's own self but to the world.

Carolyn Hsu
New York, June 16, 2018

后 记

我希望各位读者喜欢这本书,并能从中获益。这一中国文化精髓之旅应会激励中国读者为我们的文化遗产而自豪,而西方读者则有机会更充分地欣赏中国人的生活之道。

很少有文化拥有可以追溯到3000年前的社会礼仪和餐桌礼仪规范。这些源远流长的礼仪规范可能说明了一个事实:其他文明和王朝兴衰更替,而中国文化却得以传承千年。我们文化的持久性无疑可以让年轻一代中国人引以为豪,无论时代如何变迁,千百年来指引我们的这些教诲都将令所有华人得到慰藉。

中华美食是我们文化的核心组成部分,也是传播中国文化的最佳使者,在全球每个角落为人们敞开大门。我希望睿智的读者们能够思考,中国文化带来的最好影响是什么,不仅仅是对自己,更是对世界。

徐芝韵

纽约,2018年6月16日